IMAGES
of Aviation

AVIATION IN SOUTHERN OREGON

ON THE COVER: In 1933, Boeing introduced the Model 247, considered by most to be the first modern airliner. Most of the initial production run was sold to United Airlines, but the plane's popularity quickly waned with the introduction of the larger and more capable Douglas DC-3. In this image, passengers and crew pose alongside a Boeing 247 bearing the 1936 UAL logo. (SOHS No. 16791.)

IMAGES
of Aviation

AVIATION IN SOUTHERN OREGON

Bill Alley and the
Southern Oregon Historical Society

ARCADIA
PUBLISHING

ISBN 978-0-7385-8191-0

Published by Arcadia Publishing
Charleston, South Carolina

Printed in the United States of America

Library of Congress Control Number: 2010932738

For all general information, please contact Arcadia Publishing:
Telephone 843-853-2070
Fax 843-853-0044
E-mail sales@arcadiapublishing.com
For customer service and orders:
Toll-Free 1-888-313-2665

Visit us on the Internet at www.arcadiapublishing.com

For Dick Alley and George Milligan, who introduced me to flying long ago, and Aubrey Sander, who kept me writing about it.

Contents

Acknowledgments

A great many people have generously contributed to the completion of this book. Foremost, I would like to acknowledge the role of the Southern Oregon Historical Society (SOHS) for providing access to their unparalleled archival collections. Their images form the foundation of this book. SOHS's director, Allison Weiss; Kathy Enright; and volunteers such as Alice Mullally and Carol Holst contributed many hours preparing the images contained herein. I would also like thank the Rogue Valley International Airport–Medford and Ashland historian Terry Skibby for the use of their photographs.

–Bill Alley
Vancouver, Washington

Introduction

Of all the cities and towns in Oregon, Medford—situated in the southern end of the Rogue River Valley—played a dominant role in the development of aviation in the Pacific Northwest. The city's strategic location halfway between Seattle and San Francisco, coupled with local recognition of the benefits of aviation, secured Medford's preeminent place in Oregon's aviation heritage.

Like countless other cities and towns across the nation, Medford had seen its fair share of itinerant daredevil balloonists, but it was not until 1908 that southern Oregon residents first heard the roar of an airplane engine. Eugene Ely, who would later gain fame as the first man to land and fly from the deck of a naval vessel, set up his aircraft on the Cox ranch west of Medford. Ely's intention was to carry passengers, but the combination of an underpowered engine and the relatively high altitude prevented him from becoming airborne. When Ely returned to southern Oregon in 1911, flying for the Curtiss Aircraft Company, he successfully carried passengers from the city baseball field. Two years later, Medford played host to the region's first Fourth of July air meet.

In 1919, the first locally owned aircraft made its appearance over the skies of Medford. Floyd Hart and Seely Hall had purchased a surplus Curtiss JN-4 "Jenny," which they operated from the Gore Ranch. According to the *Medford Mail Tribune*, for "five and ten dollars a head" passengers were treated to a short 10-minute flight around the valley. Operating out of a stubble field, however, took a toll on the Jenny's large wooden propeller; a new site was found on the south-side end of Fir Street in Medford. Hall and Hart named their aircraft the *Mayfly* one cold morning when it seemed doubtful that the plane would fly.

The *Mayfly* had the distinction of being the first plane to carry a passenger from Medford to another destination. In August 1919, the Elks held a convention in Klamath Falls. Since both Hall and Hart were Elks, they hit upon the idea of flying the exalted ruler, George Collins, to the convention with Hart at the controls. It was hoped that the *Mayfly* could offer rides while in Klamath Falls, but the altitude made it difficult to take off with the added weight of a passenger, so Hart returned to Medford alone.

As the 1920s dawned, the U.S. Forest Service began contemplating the use of aircraft to patrol the region's forests for early fire detection. With the U.S. Army Air Service facing draconian cuts in their operating budgets, they quickly agreed to supply the planes and pilots for such a venture to keep their men trained. When the forest service selected Medford as a base of operations, the City of Medford purchased a tract of land south of town adjacent to the county fairgrounds to accommodate them. The new airfield was named Newell Barber field in honor of a young Medford man who lost his life on his first flight during World War I in Europe. Newell Barber Field became the first public airfield in the state of Oregon.

It was with the assistance and support of the federal government, in particular the U.S. Post Office, that commercial aviation was able to take firm root across the country. Post office officials had become convinced of the usefulness of a series of airmail routes across the country to facilitate rapid delivery of the mail. In 1925, Congress passed the Kelly Act, authorizing private carriers to bid on airmail routes. Among those was Contract Airmail Route (CAM) 8. CAM 8 was the longest of the proposed routes, running the length of the Pacific coast, from Los Angeles to Seattle.

Vern Gorst, a former Medford resident who had established automobile stage services in Medford and Coos Bay, recognized the potential of the fledgling airmail service. He organized the Pacific Air Transport Company (PAT) and bid on the West Coast route. To raise the necessary capital, Gorst sold shares in his company to anyone willing to invest $100 per share. In addition to acquiring the backing of financial heavyweights such as Julius Meier of the Meier and Frank department stores, Gorst even had his taxi drivers solicit investments from passengers. Seely Hall of Medford recalled selling shares on the street.

On December 31, 1925, Gorst's Pacific Air Transport was awarded the CAM 8 contract. Gorst selected the Ryan M-1 mail plane as his primary aircraft, and he and Claude Ryan surveyed the 1,100-mile route. Stops would be made at Seattle, Portland, Medford, San Francisco, Fresno, Bakersfield, and San Francisco, and later, Los Angeles.

Because Portland did not have a suitable landing field, the planes actually landed across the Columbia River at Vancouver's Pearson Field, establishing Medford's Newell Barber Field as Oregon's only airmail stop.

At 8:30 on the morning of September 15, 1926, PAT pilot Vern Bookwalter arrived at Newell Barber Field in his Travelaire with the first bags of regularly scheduled airmail. After a short ceremony, the bags, augmented by the first letters from the Medford Post Office, were transferred to a Ryan M-1 piloted by Pat Patterson and continued on their way south.

Without the subsidy of the post office department, the fledgling aviation industry would not have gotten off the ground when it did. By providing carriers with 80 percent of the revenues from airmail postage, the young industry was able to establish itself. It was only a matter of time before delivering mail would expand to encompass the transport of freight and passengers.

For the remainder of 1926, Pacific Air Transport carried mail exclusively. However the regularity of schedules, along with a now proven safety record, soon led to an interest in passenger flight. In January 1927, two passengers were carried by Pacific Air Transport. At the time, passenger service was considered on a space available basis, with priority given to the mail. Passenger service quickly became known as "riding the mail," as passengers were required to either sit on the bags or hold one on their lap.

Vern Gorst later recalled that many of the pilots resisted carrying passengers. Because the pilots were equipped with parachutes, they would not hesitate to bail out of a plane should trouble develop. With a passenger on board, however, the pilot could not very well bail out and leave his charge; he was required to remain with his ship.

The two passengers in January blossomed into a massive demand for passenger service—by August 1927, Pacific Air Transport had carried 216 passengers. Unlike most of the major carriers across the country, however, Gorst's Pacific Air Transport was severely underfinanced. Pilots and fuel bills were often paid in stock, and the occasional loss of an airplane placed enormous financial burdens on the company.

In late 1927, Gorst sold his interest in Pacific Air Transport to the Boeing Air Transport Company, which operated the San Francisco–Chicago airmail route and enjoyed a sound financial foundation. A new holding company was formed, the Boeing Airplane and Transport Company, which consisted of the Boeing Aircraft Company, Boeing Air Transport, and Pacific Air Transport. The following year, Pratt and Whitney—manufacturers of aircraft engines—was included and the company reorganized as the United Aircraft and Transport Company. With the acquisition of National Airlines and Varney Airlines in 1930, the company would again be reorganized, this time as United Airlines.

In 1928, a new air carrier named West Coast Air Transport announced that it would begin regular service to Medford's Newell Barber Field. West Coast Air Transport was a subsidiary of the Pickwick Stage System and operated a Bach Air Yacht trimotored aircraft. Boeing, as operator of Pacific Air Transport, also planned to place larger aircraft on their route and introduced the Boeing 40-B. This new aircraft had an enclosed cabin with seating for up to four passengers; no longer would patrons be required to "ride the mail." Boeing and PAT would also acquire the four-passenger Fokker Universal to ensure customer comfort.

The introduction of larger, more capable aircraft, however, created new problems. Newell Barber Field proved to be inadequate to handle the larger, new aircraft coming in to service. In March 1928, West Coast Air Transport announced that it would temporarily discontinue its Medford service in favor of Eugene, Oregon. Several days earlier, West Coast Air Transport's ship had been so mired in the old field that several passengers had to be removed and sent on by train to lighten the load for take off.

The Ford Reliability Air Tours had their origins in 1925 as a means to showcase the reliability and safety of commercial aviation. The fourth tour, in 1928, brought the national event to the far western states, and included a stop at Medford's Newell Barber Field. Medford was the smallest field on the air tour itinerary. While the pilots publicly praised the Medford field, as they undoubtedly did for all the airports they visited, there was an undercurrent of dissatisfaction with the field and its amenities, especially by the pilots of the larger Ford trimotor planes. This air tour, along with so many other events in Medford's aviation heritage, was captured on film by Horace Bromley for the California Oregon Power Company (COPCO) current events newsreels. These invaluable moving images are now preserved in the archives of the Southern Oregon Historical Society.

The biggest threat to Medford's position as one of the premiere aviation cities in Oregon came on the heels of West Coast Air Transport's decision to discontinue its Medford service. The army and navy departments, and the aviation insurance underwriters, condemned Newell Barber Field as a destination for the new, large trimotored aircraft being acquired by most carries. The field's location, between the Jackson County fairgrounds and Bear Creek, made it impossible to undergo expansion; the rapid advancements in aviation technology and design had rendered the field obsolete and ensured its demise after a mere decade of service.

In response to this threat, the Medford Chamber of Commerce formed an aviation committee to work closely with commercial, local, and federal aviation interests to locate a site for a new, larger airfield. This committee ultimately settled upon a site off Biddle Road, three miles north of downtown. In order to finance the construction of a new airport, it was necessary for the City to place a bond measure on the ballot, scheduled for a vote on April 2, 1928. The measure called for the issuing of $120,000 in bonds for construction of a new, modern facility, including a modern terminal and runway. Should the voters of Medford approve the measure and construct a Class A airport, the department of commerce agreed to designate Medford as a terminal airfield and locate a $50,000 regional communication station and a 2.5 million candlepower beacon at the new facility.

The citizens of Medford approved the airport measure overwhelmingly, and construction began on the new runway, an expansive 3,600 feet long by 120 feet wide and surfaced with gravel that would easily accommodate even the largest aircraft of the day. It was a significant improvement over Newell Barber Field, and there was sufficient surrounding acreage for future expansion. The new runway was to be illuminated with 75 border lights around the perimeter to allow for night landings. In addition to the improved airstrip, the new facility featured a modern administration building with room for a coffee shop, radio room, and passenger amenities and a large attached hangar designed by noted Medford architect Frank C. Clark. Six different oil companies contracted to provide high octane aviation fuel. Pacific Air Transport and others began using the new Medford Municipal Airport in 1929, although the official dedication did not take place until 1930.

In spite of the dramatic increase in passenger traffic, it was still the lucrative airmail contracts that provided the sound financial foundation of most airlines. The financial underpinnings of the entire industry underwent a dramatic shift in 1934. Under the direction of President Roosevelt, the Post Office Department cancelled all the airmail contracts and turned the responsibility over to the U.S. Army Air Corps. The results were disastrous and the loss in untrained airmail pilots staggering.

With the loss of the airmail subsidies, there was a dramatic shake out among the many carriers established during the 1920s. Only those able to subsist on passenger traffic were able to survive.

The government eventually renewed the airmail contracts but forced the breakup of the large conglomerates, prohibiting the mergers of manufacturers and carriers. This forced the reorganization of PAT's parent company, now separated from the manufactures Boeing and Pratt and Whitney, into United Airlines, putting an end to Pacific Air Transport's operation as a separate entity.

Even with the resumption of the airmail contracts, it was now passenger service that had become the mainstay of the commercial aviation industry. New, faster ships capable of carrying increasingly larger numbers of passengers had made the airlines profitable.

Throughout the 1930s, the sight of military aircraft at the Medford Municipal Airport was not uncommon. Just as was the case with the airmail and early passenger flights, Medford's location made it an ideal stopover for refueling aircraft and resting flight crews. As tensions grew in both Europe and Asia, the U.S. Army Air Corps (USAAC) began stepping up its training flights, and Medford experienced increasing activity. In July 1941, about 135 USAAC planes passed through Medford's airport and an even greater number did in August. To accommodate this traffic, a permanent military presence was established with the stationing of a "service detachment" to provide maintenance for aircraft in transit. Even a military canteen was set up in the old commerce department radio shack.

Shortly after America's entry into World War II, the City of Medford signed a lease with the U.S. Army Air Corps, turning the airport over to the military for the duration of the war plus six months; the cost was $1 per year. With the exception of the regularly scheduled United Airlines flights, the Medford Municipal Airport was henceforth a military installation and off limits to civilians.

The army made a number of improvements to the airport during its tenure. In 1943, it established the West Coast's first approach control, providing an onsite traffic control system. The army also acquired several additional parcels of land. These additions complicated the eventual transfer of the airport back to the City in 1948.

In 1949, Medford automobile dealer Art Winetrout was stricken with polio and rushed to medical facilities in Portland. The trip by ambulance took more than 12 hours, and the patient died four days later. For George Milligan, the young civil aviation administration tower man (better known today as an air traffic controller) at the Medford airport, this tragedy provided another example of the Rogue Valley's dire need for an air ambulance service. Milligan, admiringly described by *Medford Mail Tribune* editor Eric Allen as "an Irishman who is part Wrong-Way Corrigan and part Florence Nightingale," set out to make his dream a reality.

Mercy Flights, Inc., the country's first nonprofit air ambulance service, was organized in August, and Milligan began pitching his idea to anyone who would listen. And he found a ready audience. Milligan's vision was to provide at-cost medical transport for anyone who could afford it, and at no cost to those who could not. By the end of the year, enough money had been raised—almost a third of it from schoolchildren—to purchase and equip a surplus twin engine Cessna. Milligan and several other qualified pilots signed on to fly the plane for free, and local mechanics offered to donate their time for the plane's maintenance. The City of Medford, which operated the airport, provided tie down space for the plane at no cost.

For the next 20 years, Mercy Flights and its good Samaritan flight and ground crews transported a steady stream of patients. Their growing fleet of aircraft, emblazoned with the American Red Cross symbol, was the only civilian aircraft authorized to display this international symbol. On many occasions, services were required in the midst of dramatically inclement conditions. The log books of Mercy Flights abound with tales of dramatic rescues in remote locales under weather conditions that would deter many other courageous pilots.

Over the years, the Medford Municipal Airport has enjoyed a number of expansions, upgrades, and improvements, maintaining its position as one of Oregon's oldest and most active airports. With the opening of the Free Trade Zone, the airport was upgraded to international status, and a public contest was held to rename the facility; the winning entry was Rogue Valley International-Medford Airport. The airport currently services approximately 600,000 passengers a year and has welcomed aircraft as large as the Air Force One 747 and the even larger Russian Antonov 124.

One

BALLOONISTS, BARNSTORMERS, AND THE *MAYFLY*

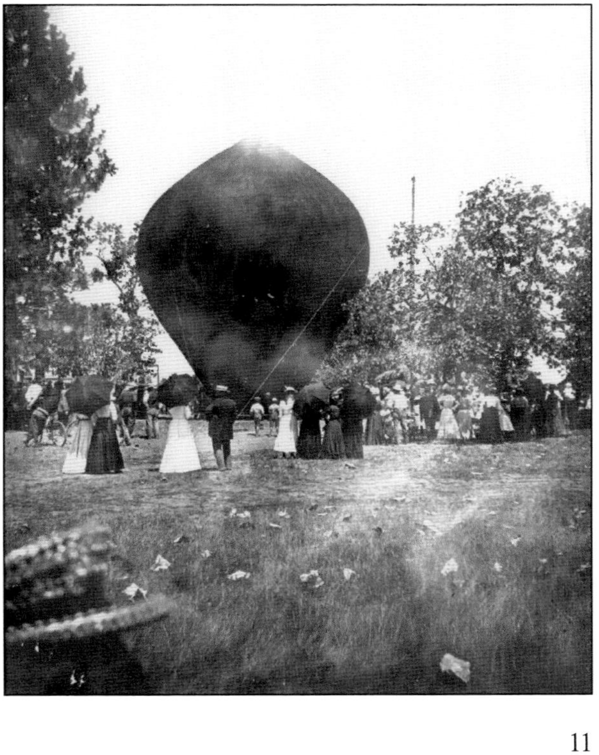

"Professor" Charles Nelson, a travelling balloonist, arrived in Medford to perform on the Fourth of July, but unfavorable winds postponed his exhibition until July 8. On that day, a large crowd gathered at what is now the site of the Carnegie Library to watch as Nelson, suspended from a trapeze below his balloon, rose to 2,200 feet before parachuting to the ground. The deflated balloon followed him down. (SOHS No. 1550.)

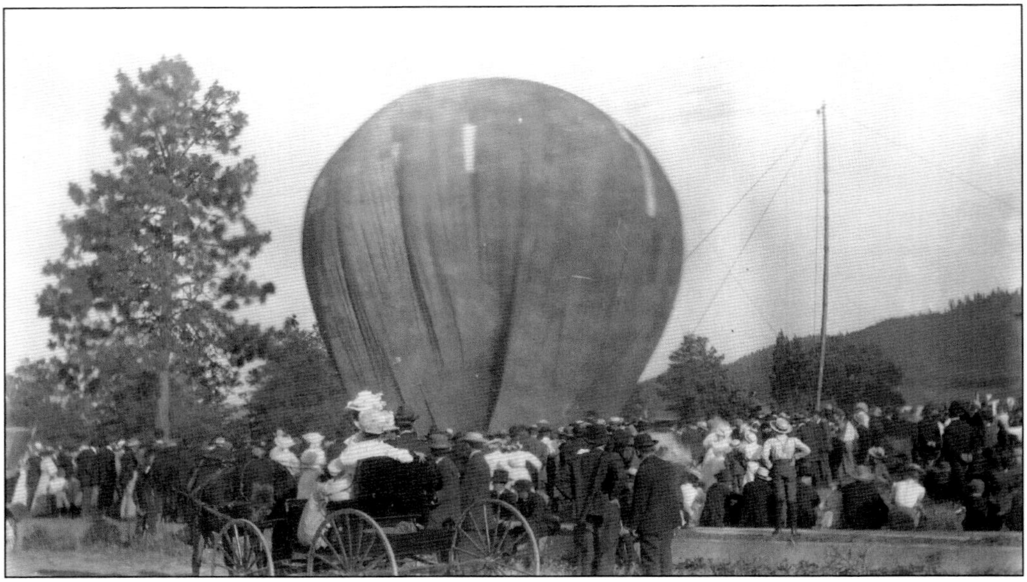

Balloonists other than Nelson made visits to the Rogue Valley. They traveled in by train, held an exhibition, then moved on to the next town. (SOHS No. 12086.)

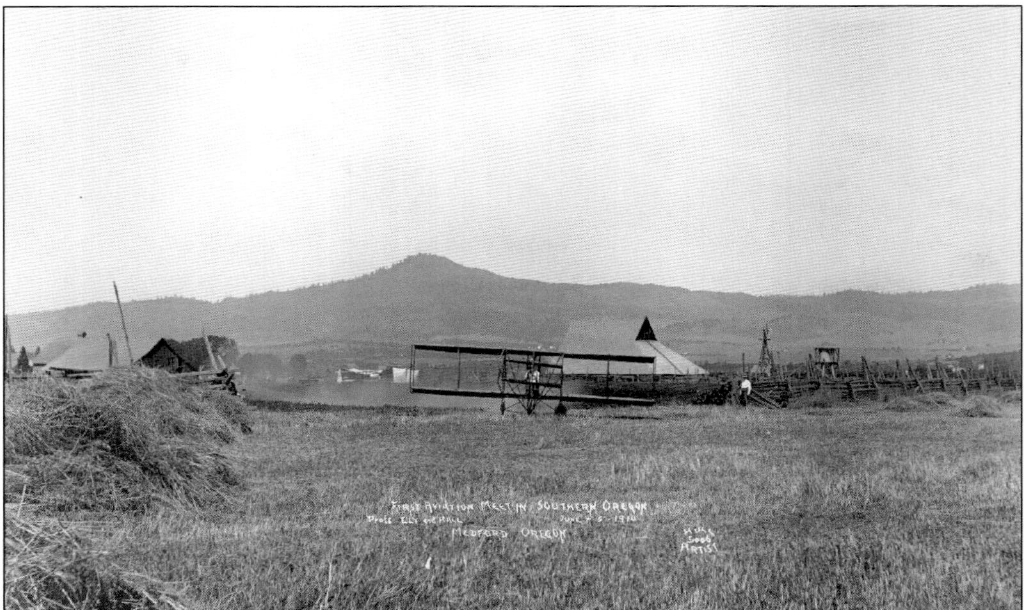

Medford residents first heard the sound of an aircraft engine in 1908, when Eugene Ely came to town for a flying exhibition. The combination of an underpowered motor and relatively high altitude, however, prevented him from carrying any passengers aloft and Ely had to refund the money he collected. He would have better luck on his next visit. (SOHS No. 1.)

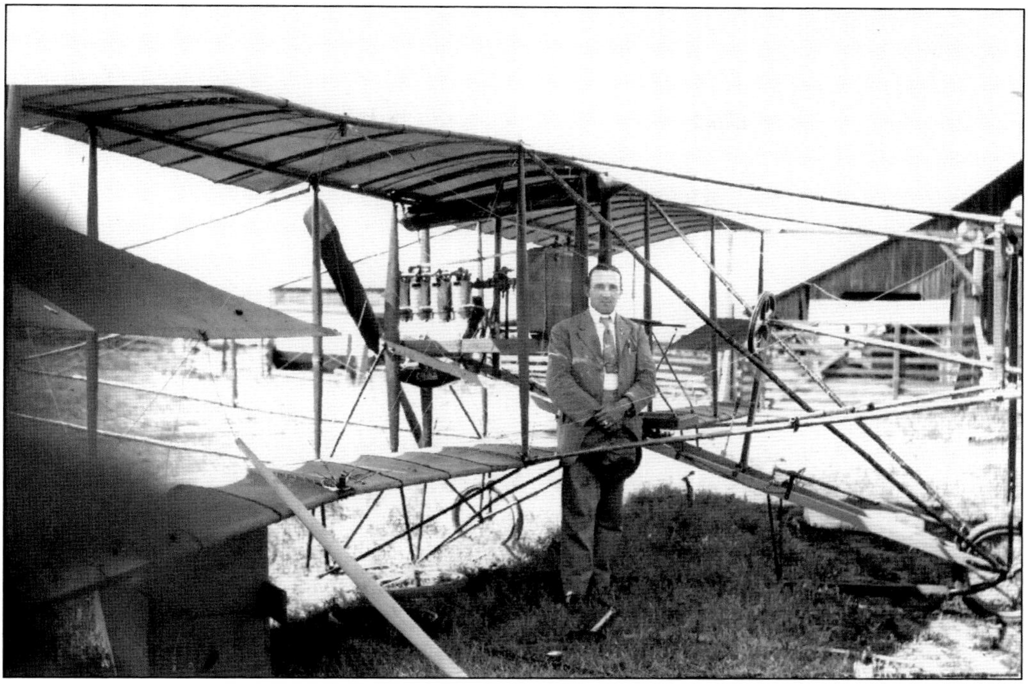

Eugene Ely became famous as the first man to fly off a naval vessel in the fall of 1910. Earlier that same year, Ely returned to Medford and successfully dazzled spectators with his flying skills in a Curtiss pusher. Ely was killed on October 19, 1911, when he crashed during an exhibition in Mobile, Alabama. (SOHS No. 15263.)

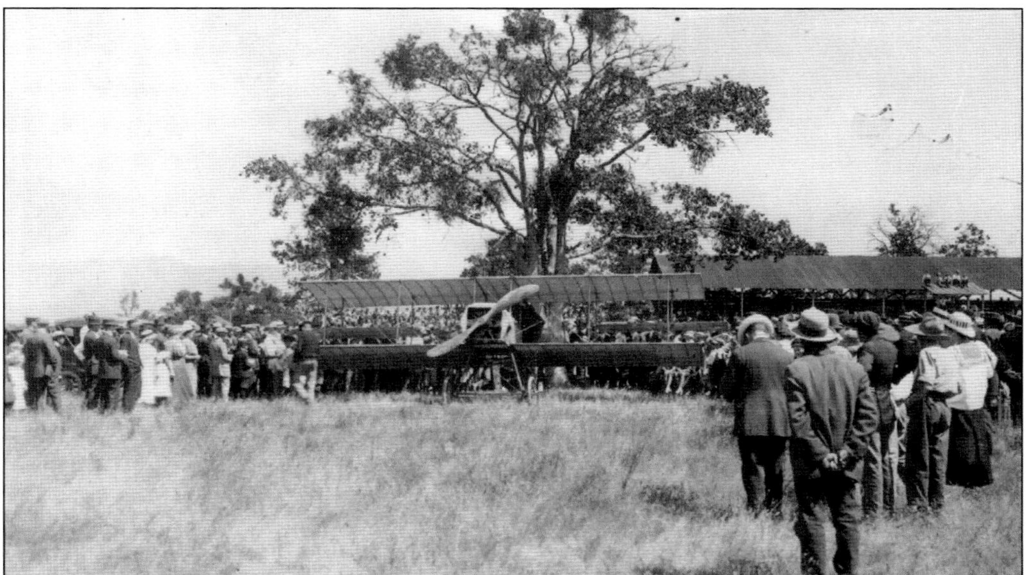

Fred De Kor was a famous acrobatic pilot who entertained crowds across the county in the second decade of the new century. In June 1915, the aerialist set his plane up at the Jackson County Fairgrounds for a demonstration of his skills. He advertised "looping the loop guaranteed or no pay." (SOHS No. 3.)

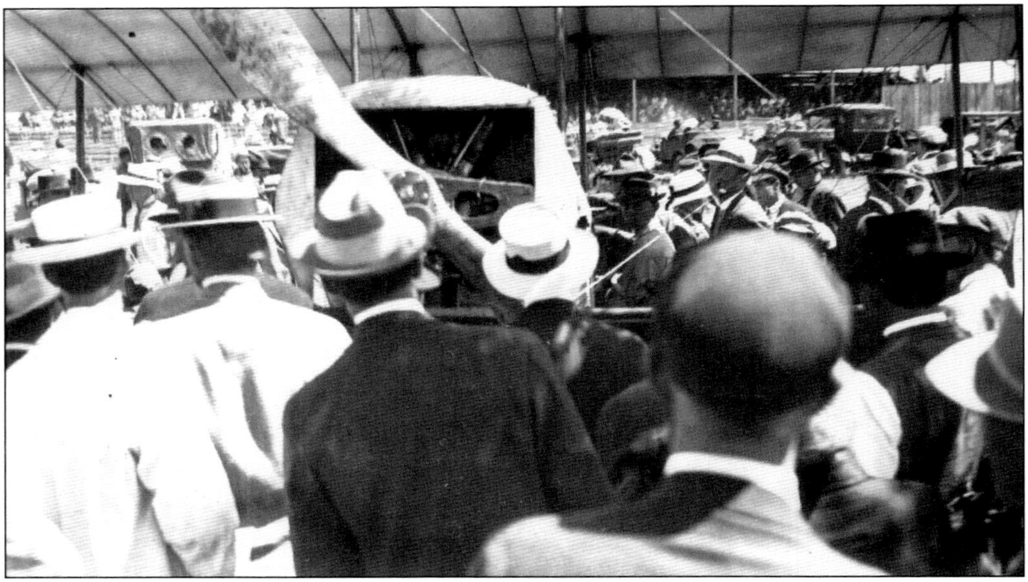

De Kor gave southern Oregon crowds much more than they bargained for on June 21. After the completion of his aerial stunts, the pilot made his approach to land when his engine failed. Uninjured, the pilot was reported to have voiced "opinions not permissible to print." The plane was a complete wreck, and the crowds swarmed the site and stripped the craft for souvenirs. (SOHS No. 5.)

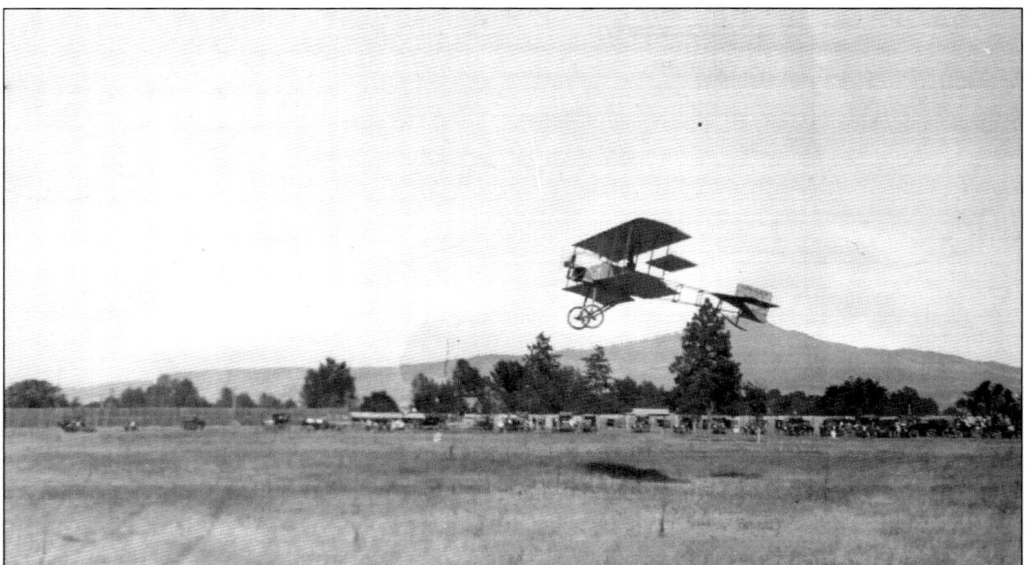

Fred De Kor takes to the skies from the fairgrounds on June 21, 1915. The automobiles in the background give an indication of the size of the crowd. (SOHS No. 2.)

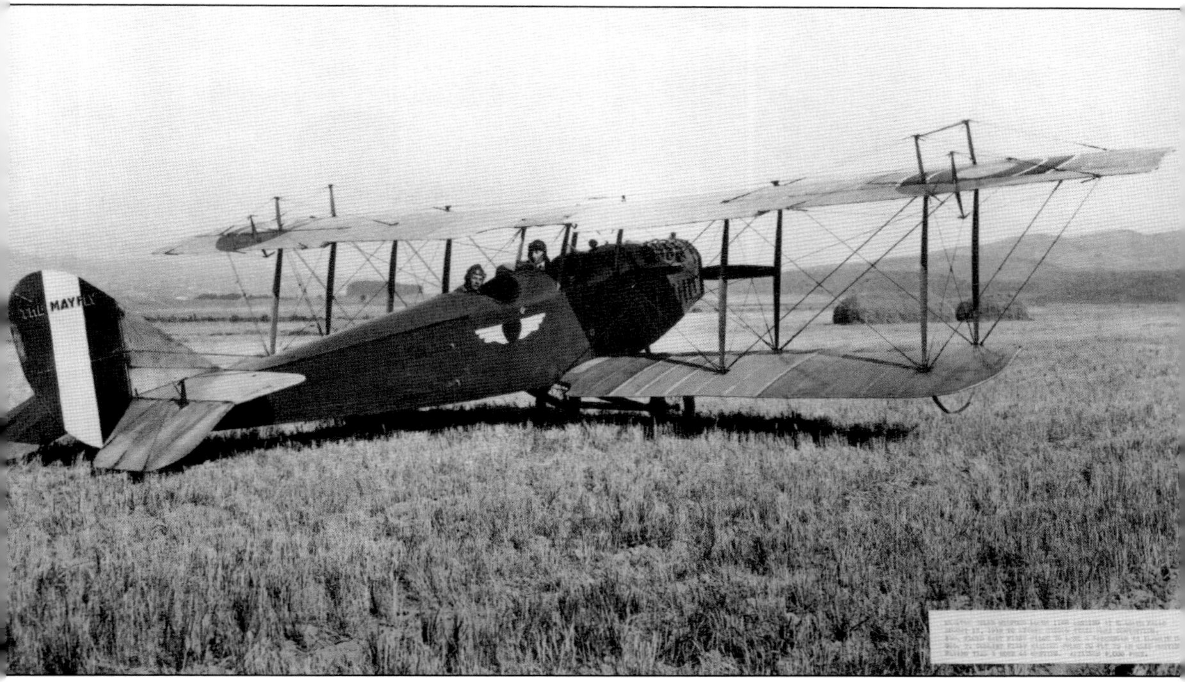

During the summer of 1919, the Benevolent and Protective Order of the Elks held a convention in Klamath Falls, Oregon. With Floyd Hart at the controls of the *Mayfly*, grand ruler George Collins was flown from Medford to Klamath Falls. Hart and Hall had hoped to raise some money by providing rides at $10 per person, but the high altitude and the underpowered OX-5 engine simply were not up to the task. Hart had to return to Medford by himself. (SOHS No. 12490.)

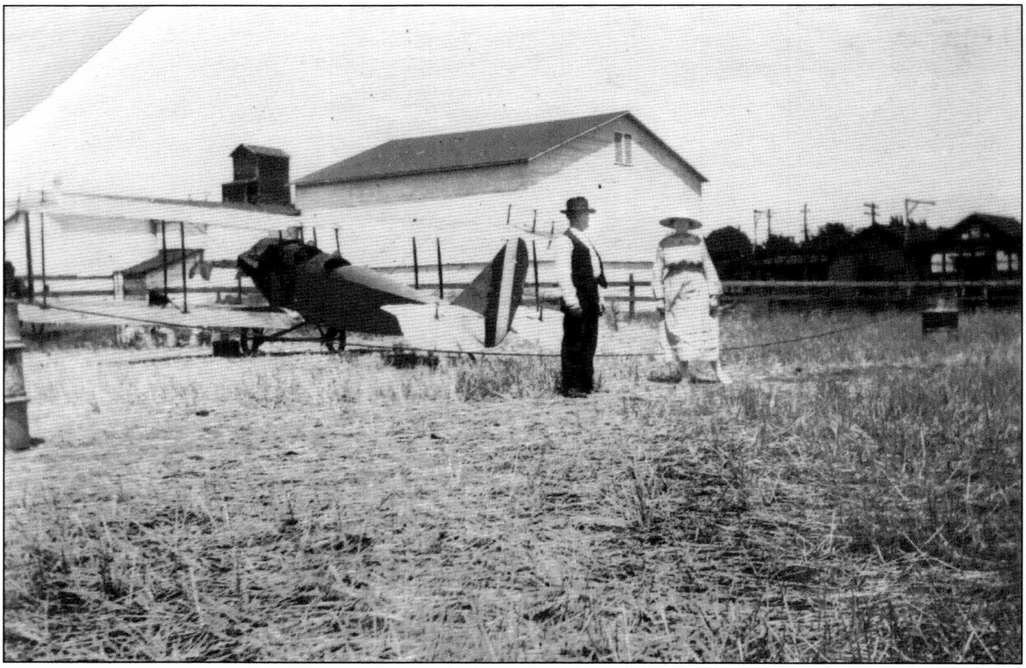

In 1919, Floyd Hart and Seely Hall purchased Medford's first home-based aircraft, a surplus Jenny. They dubbed the plane "Mayfly" because such was the plane's temperament. The Mayfly operated out of stubble fields on area farms. Here the *Mayfly*, roped off to keep onlookers at a distance, sits on the field of a farm in Central Point, undoubtedly to provide flights for passengers. (SOHS No. 20736.)

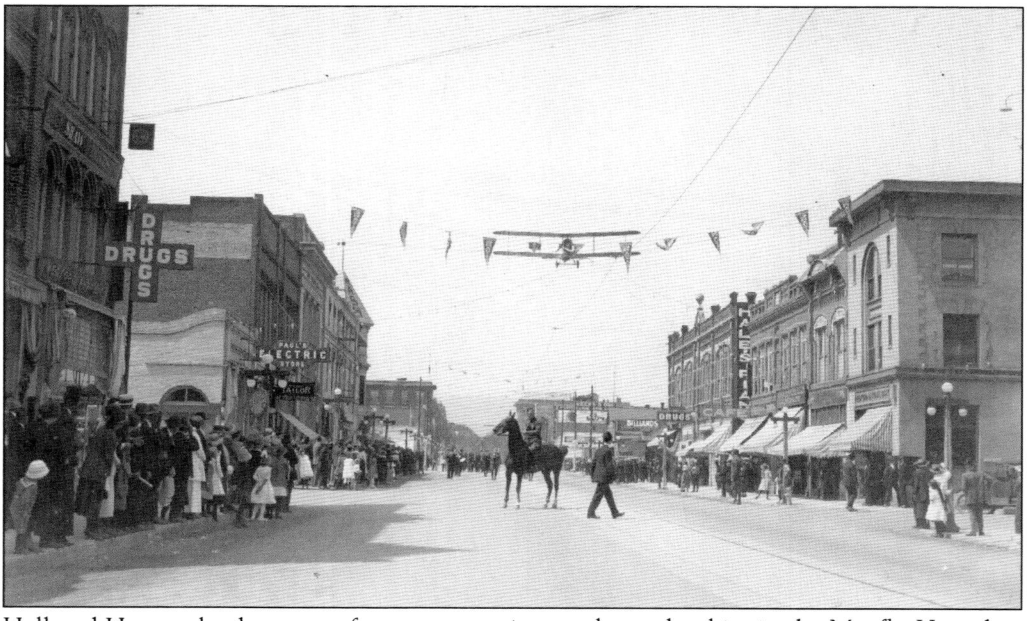

Hall and Hart took advantage of any opportunity to take to the skies in the *Mayfly*. Here they are flying over a 1919 patriotic parade in the heart of downtown Medford, at the intersection of Main Street and Central Avenue. (SOHS No. 7246.)

Two

NEWELL BARBER FIELD

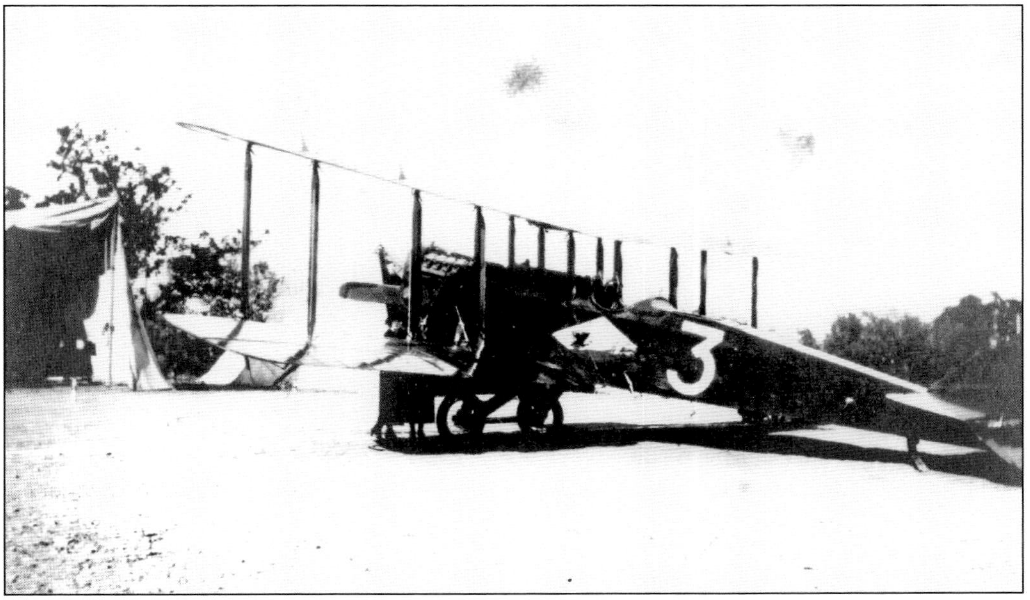

The City of Medford established Oregon's first publicly owned airfield in 1920. Named Newell Barber Field, the purpose was to provide a location for the U.S. Army Air Service and the U.S. Forest Service to operate forest fire air patrols over the forests of southern Oregon. The aircraft used were the staples of the postwar air service, the De Havilland DH-4 with a 12-cylinder Liberty motor. (SOHS No. 16333.)

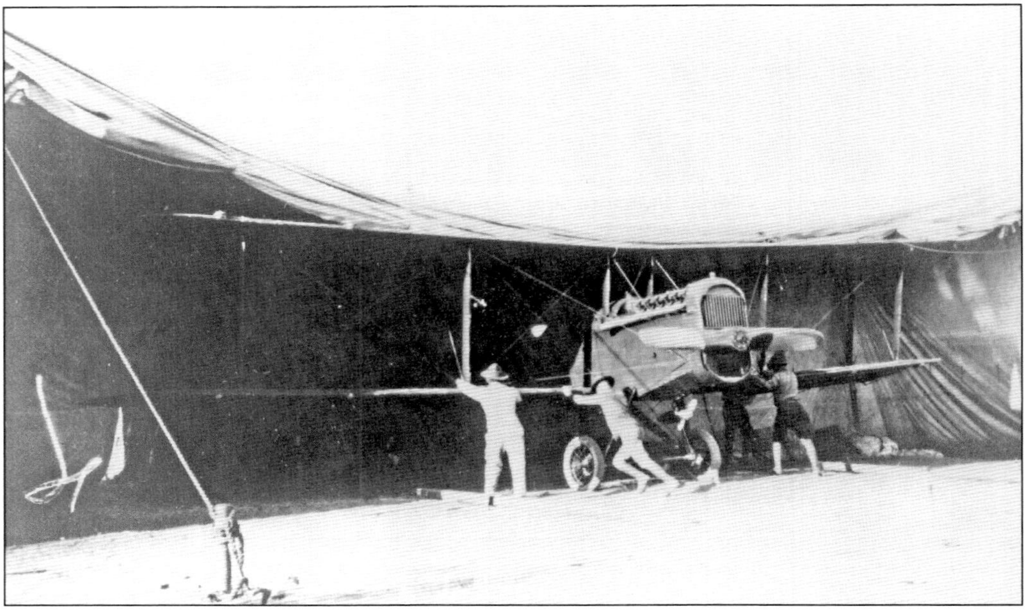

Conditions at Newell Barber Field during the fire patrol era were extremely primitive. A nearby farmhouse was used as the headquarters, and the aircraft were stored in makeshift canvas tent hangars. (SOHS No. 16334.)

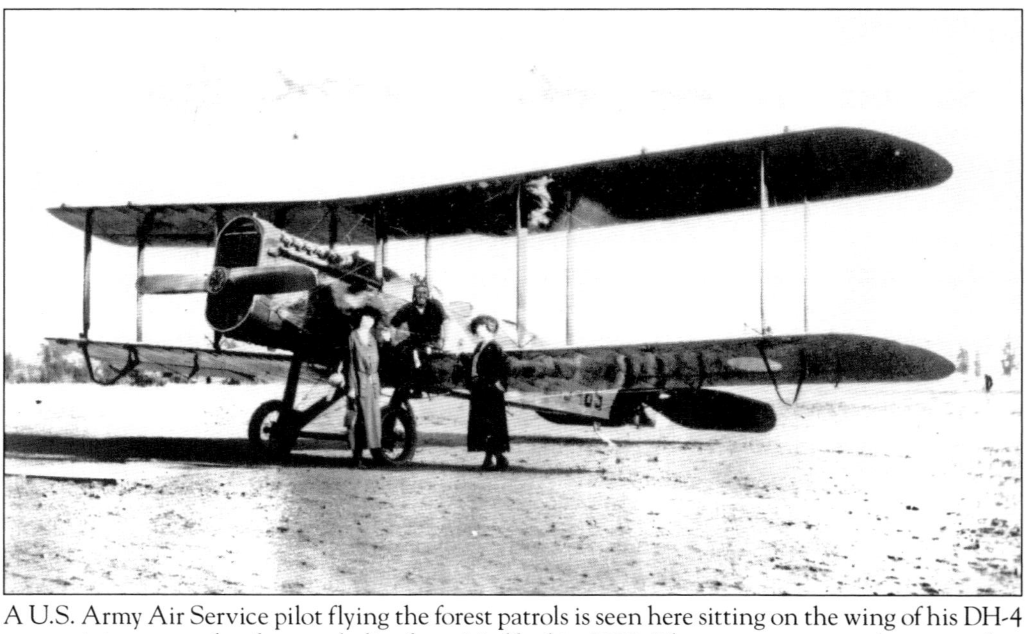

A U.S. Army Air Service pilot flying the forest patrols is seen here sitting on the wing of his DH-4 entertaining a couple of young ladies from Medford in 1920. These young army pilots proved to be a rambunctious lot and the Medford City Council was forced to ban their constant stunting over downtown. (SOHS No. 16335.)

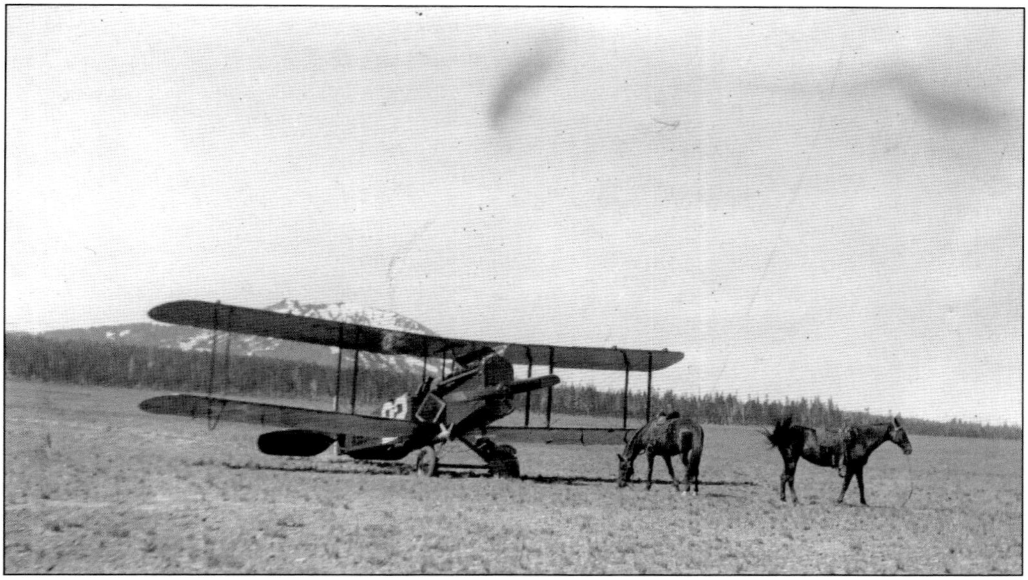

Cadet pilot Raymond Fisher and his observer, R. M. Parsons, were on a routine fire patrol out of Newell Barber Field on July 14, 1921, when the engine on their de Haviland DH-4 "started to act badly." Fisher brought his plane down approximately 5 miles northwest of Crater Lake, near Oasis Butte. It was determined that the problem was caused by faulty spark plugs, so a replacement set was flown up and dropped by parachute. (SOHS No. 15646.)

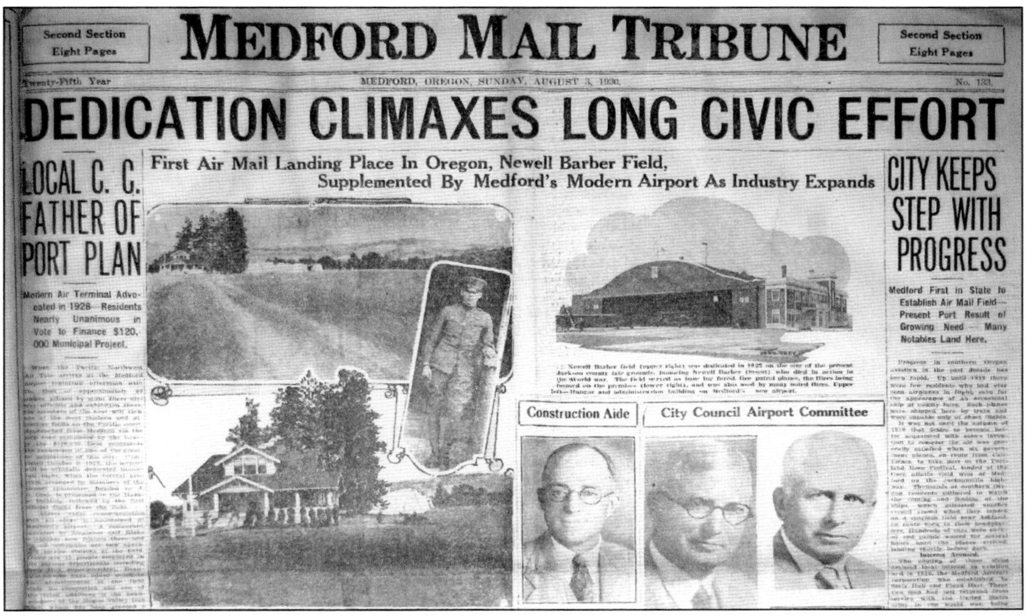

Second Section Eight Pages

MEDFORD MAIL TRIBUNE

Second Section Eight Pages

Twenty-Fifth Year — MEDFORD, OREGON, SUNDAY, AUGUST 3, 1930. — No. 133.

DEDICATION CLIMAXES LONG CIVIC EFFORT

LOCAL C. C. FATHER OF PORT PLAN

Modern Air Terminal Advocated in 1928—Residents Nearly Unanimous in Vote to Finance $120,000 Municipal Project.

First Air Mail Landing Place In Oregon, Newell Barber Field, Supplemented By Medford's Modern Airport As Industry Expands

CITY KEEPS STEP WITH PROGRESS

Medford First in State to Establish Air Mail Field—Present Port Result of Growing Need—Many Notables Land Here.

Construction Aide

City Council Airport Committee

Newspaper coverage of the dedication of the new Municipal Airport in 1930 also provided two of the earliest known photographs of the old Newell Barber Field, the farmhouse used by the army fire patrol, and the only known photograph of Medford High School graduate Newell Barber, the World War I pilot for whom the field was named. (*Medford Mail Tribune*, August 3, 1930.)

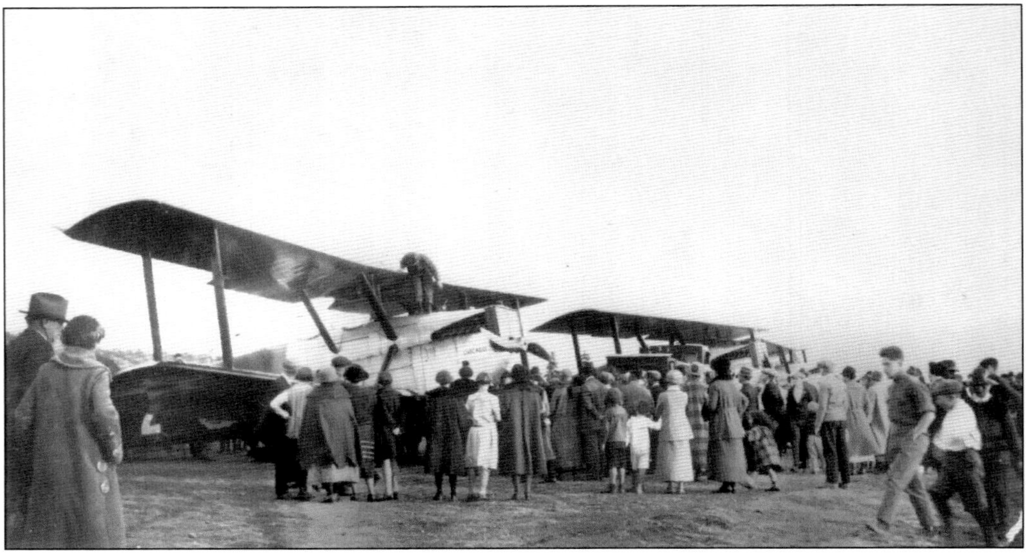

In 1924, the U.S. Army Air Service sent four Douglas World Cruiser aircraft on an attempt to be the first to circumnavigate the globe by plane. In October, two of the four completed the historic flight, returning to their starting point in Seattle. They had flown 27,000 miles in 150 days to set their record. On October 20, the World Flyers, accompanied by a support plane carrying a young journalist named Lowell Thomas, arrived in Medford on their way back to California. They were feted and treated to the opening night festivities of Medford's luxurious new Craterian Theater. (SOHS No. 15397.)

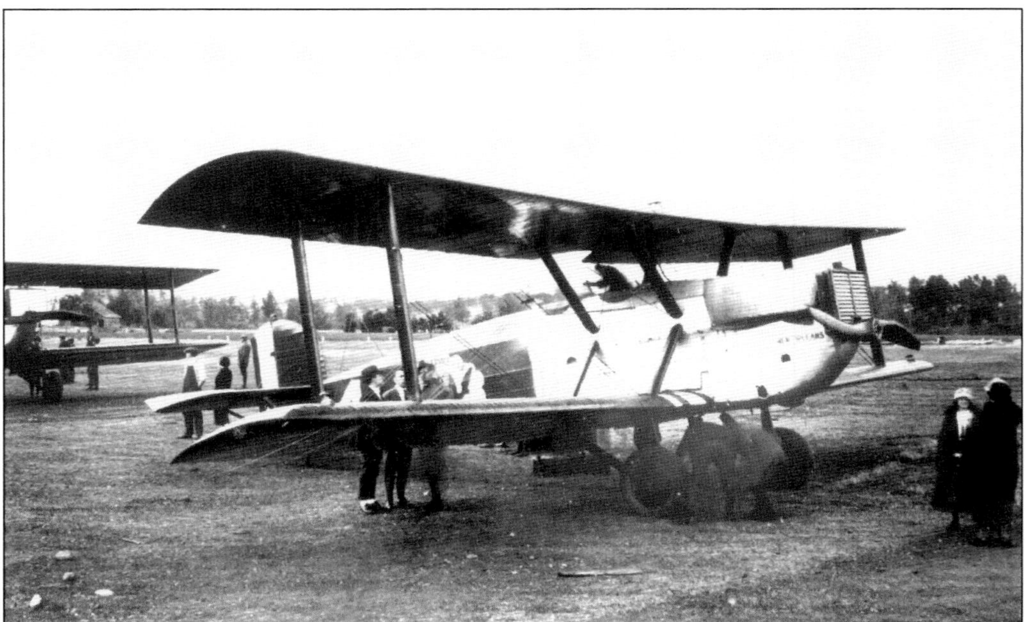

The U.S. Army Air Service World Flyers *New Orleans* and *Chicago* were the two ships to successfully circumnavigate the globe. The *Seattle* crashed in Alaska and the *Boston* went down in the north Atlantic tantalizingly close to the North American coast. Both pilots survived. Curious onlookers examine the *New Orleans* at Newell Barber Field. (SOHS No. 16682.)

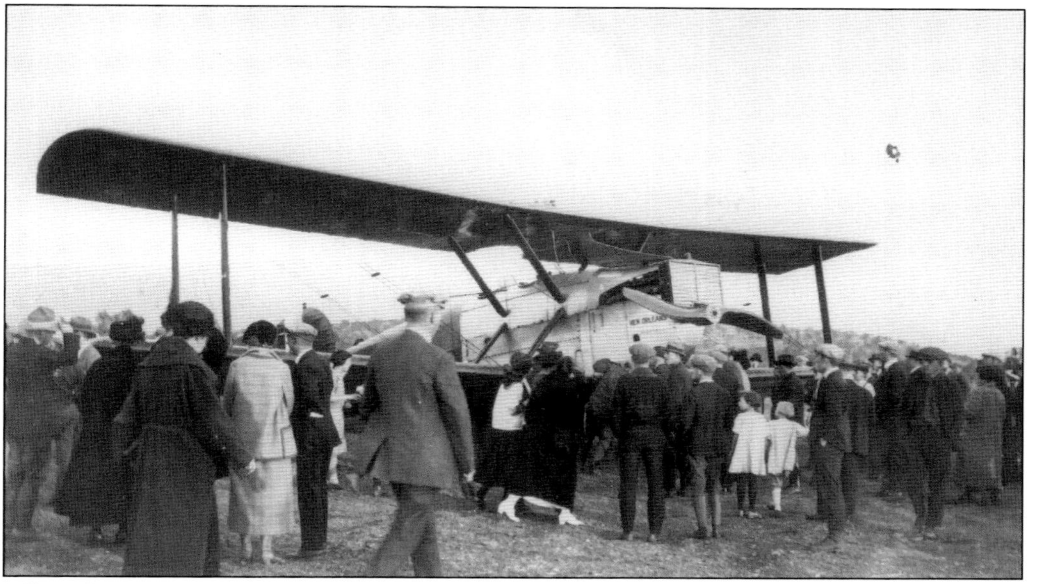

When the U.S. Army Air Service's famed World Flyers left Eugene for Medford in October 1924, word quickly spread across southern Oregon. Hundreds of enthusiastic spectators were on hand to see the famous pilots and their aircraft at Newell Barber Field. (SOHS No. 15444.)

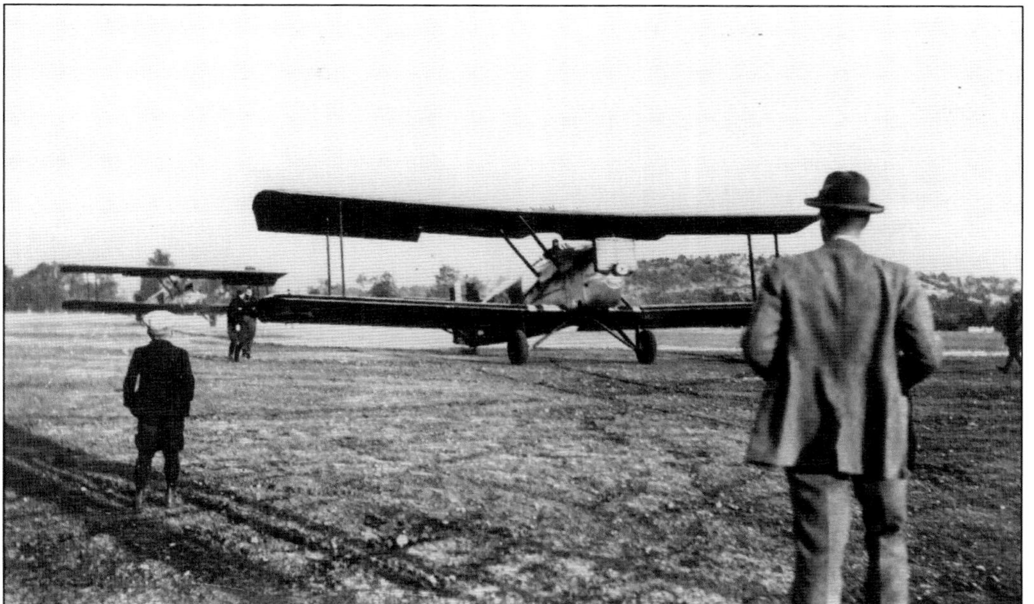

After an overnight stay at Newel Barber Field, the World Flyers fired up their 12-cylinder Liberty engines for the trip home. The little squadron was comprised of five aircraft, the two successful world flyers, the back up ship, and two support planes. Onboard one of the support planes was a young journalist named Lowell Thomas. (SOHS No. 9204.)

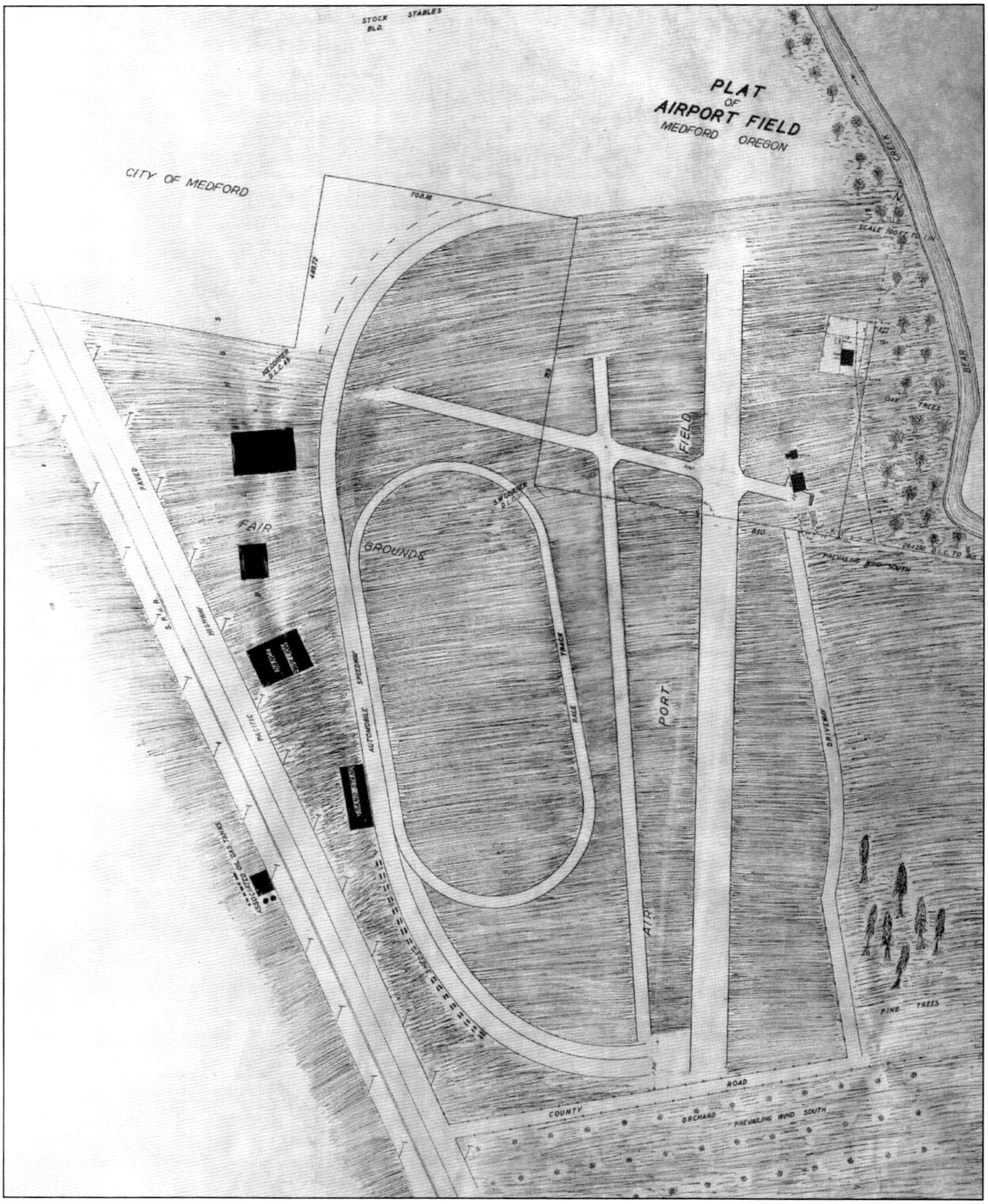

This plat of the Municipal Airfield, from around 1926, shows the close proximity that Newell Barber Field shared with the county fairgrounds. The automobile racetrack actually encircled the airfield. The Pacific Air Transport hangar and office is just east of the junction of the two strips. (Rogue Valley International-Medford Airport.)

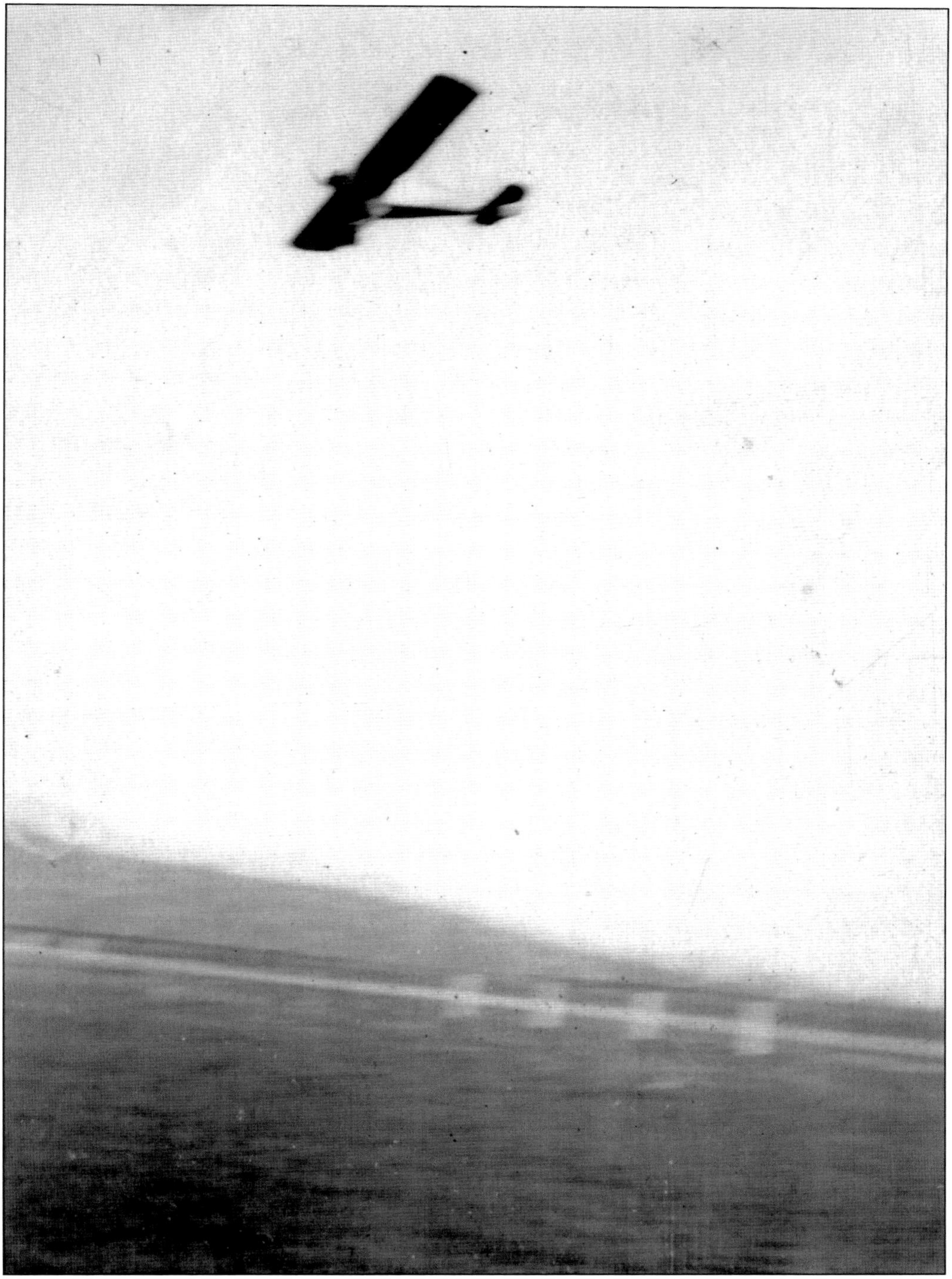

Charles Lindbergh undertook a nationwide tour after his milestone flight across the Atlantic in 1927. Sponsored by Harry Guggenheim, Lindbergh piloted his *Spirit of St. Louis* to 48 states and 92 cities. Medford was not one of the cities visited, but the "Lone Eagle" did make a fly-over, caught in this snapshot, and a message of greeting was dropped from his plane. (SOHS No. 20645.)

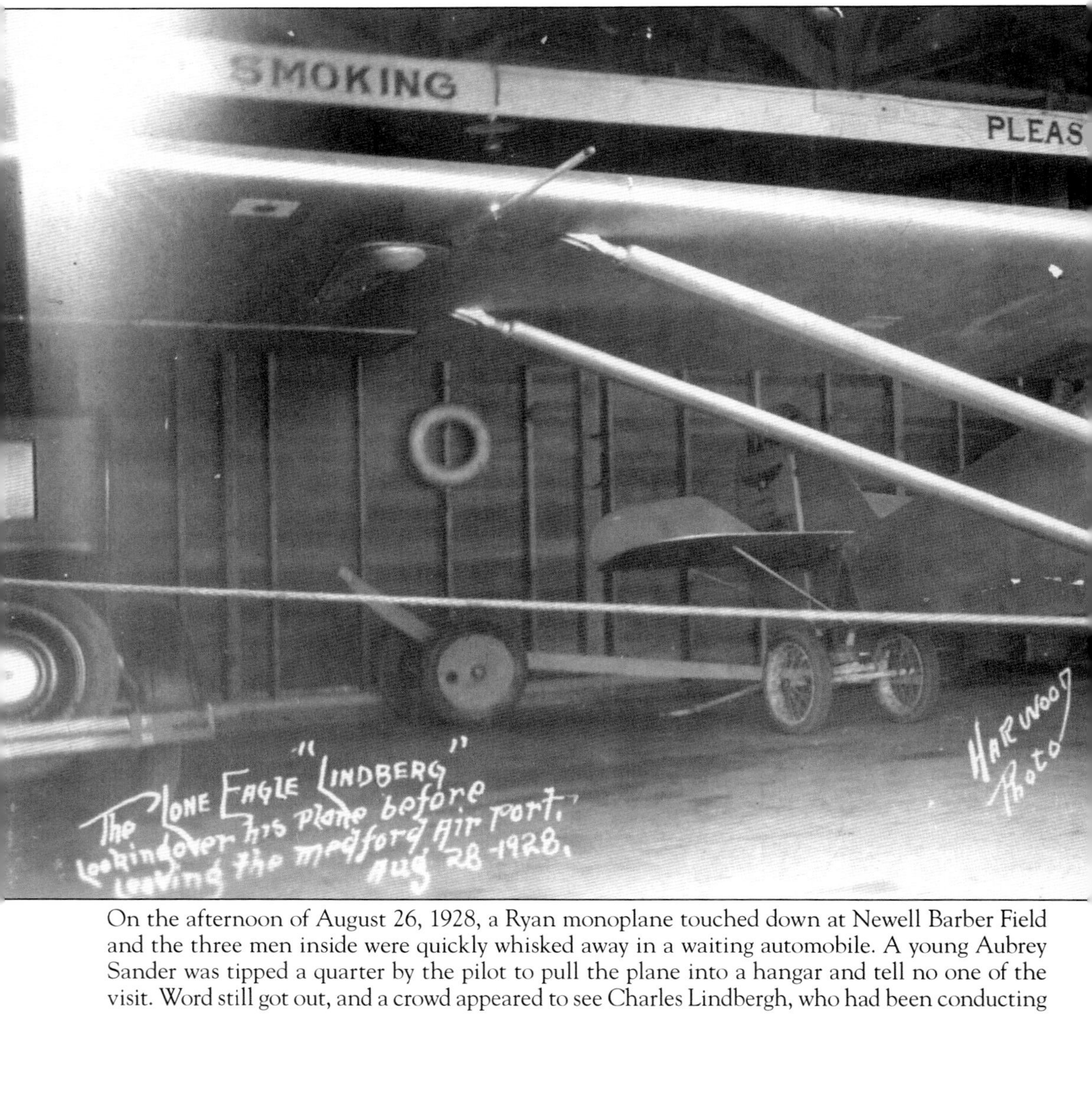

On the afternoon of August 26, 1928, a Ryan monoplane touched down at Newell Barber Field and the three men inside were quickly whisked away in a waiting automobile. A young Aubrey Sander was tipped a quarter by the pilot to pull the plane into a hangar and tell no one of the visit. Word still got out, and a crowd appeared to see Charles Lindbergh, who had been conducting

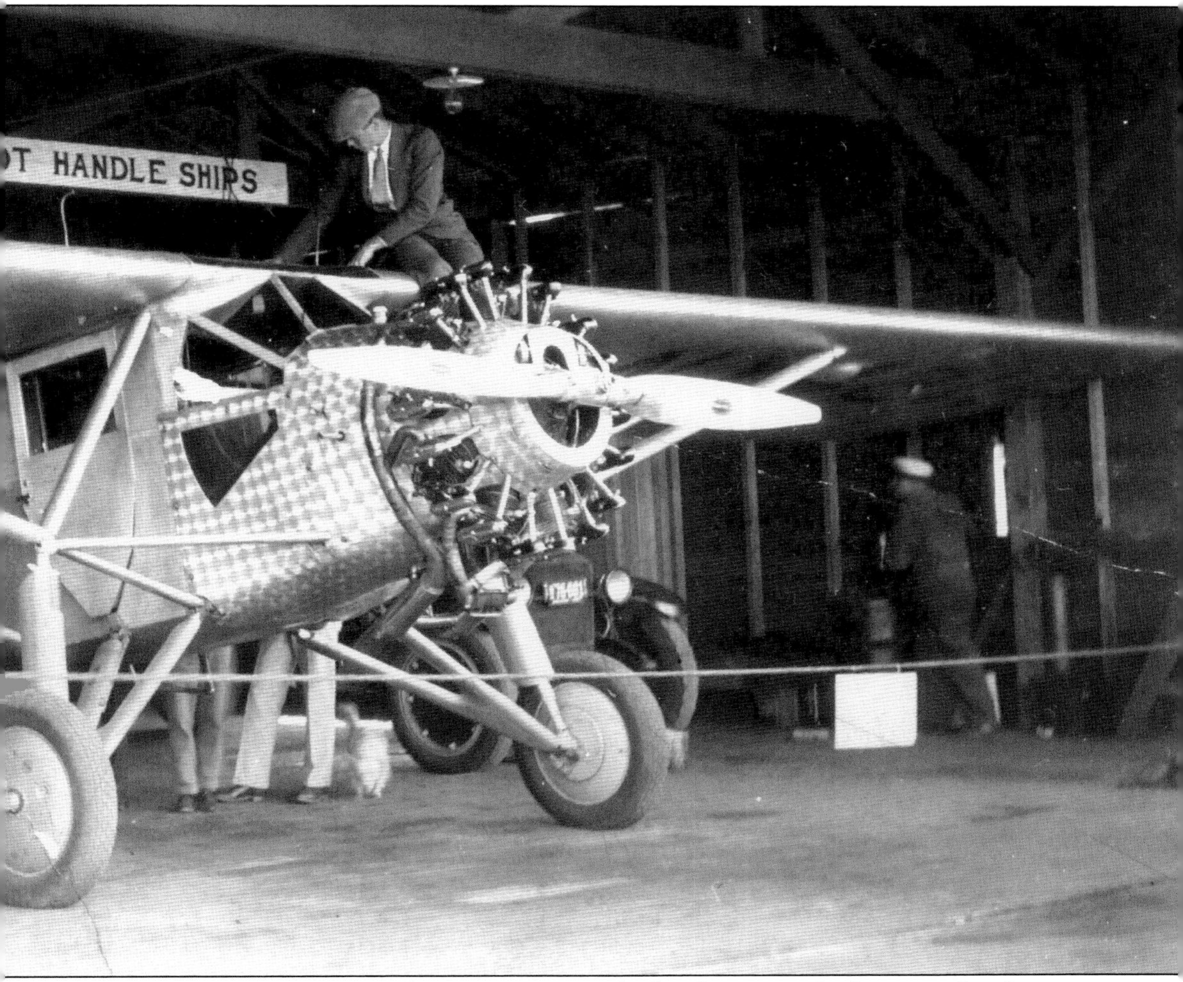

business meetings in a cabin on the Rogue River, make his departure. As for young Sander, in 1999 he jokingly told the author that enough time had passed so he could, indeed, reveal that Lindberg had landed. (SOHS No. 10423.)

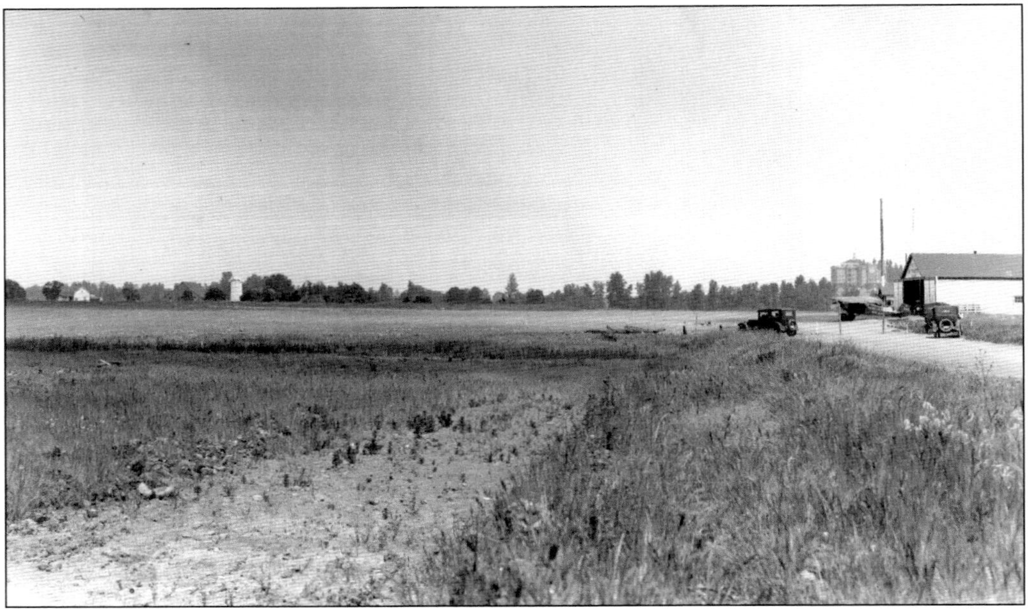

Pictured is Newell Barber Field looking north. To the right is the Pacific Air Transport hangar with a Ryan M-1 mail plane in front. In the background is Medford's Sacred Heart Hospital. The view today is decidedly different. The site of Newell Barber Field is now the South Gateway Shopping Center, and the old Sacred Heart Hospital was razed in 1966. (SOHS No. 19201.)

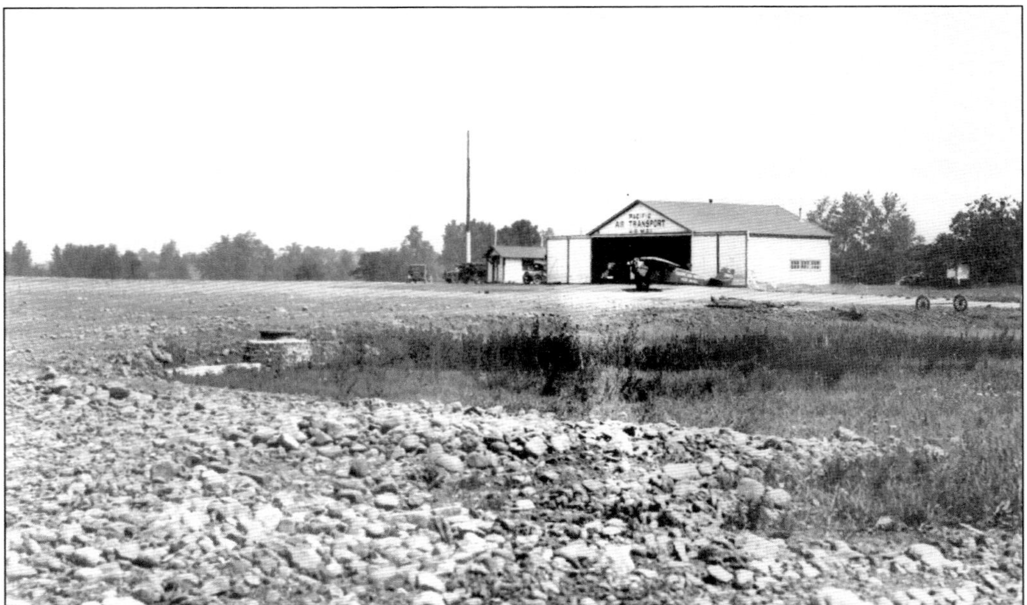

The trees and shrubs behind the Pacific Air Transport hangar and office line the course of Bear Creek. Wedged between the creek and the Jackson County Fairgrounds, Newell Barber Field had no room for expansion. This proved a problem as the decade of the 1920s came to a close and innovations in aircraft design and capabilities threatened to render the old field obsolete. (SOHS No. 19202.)

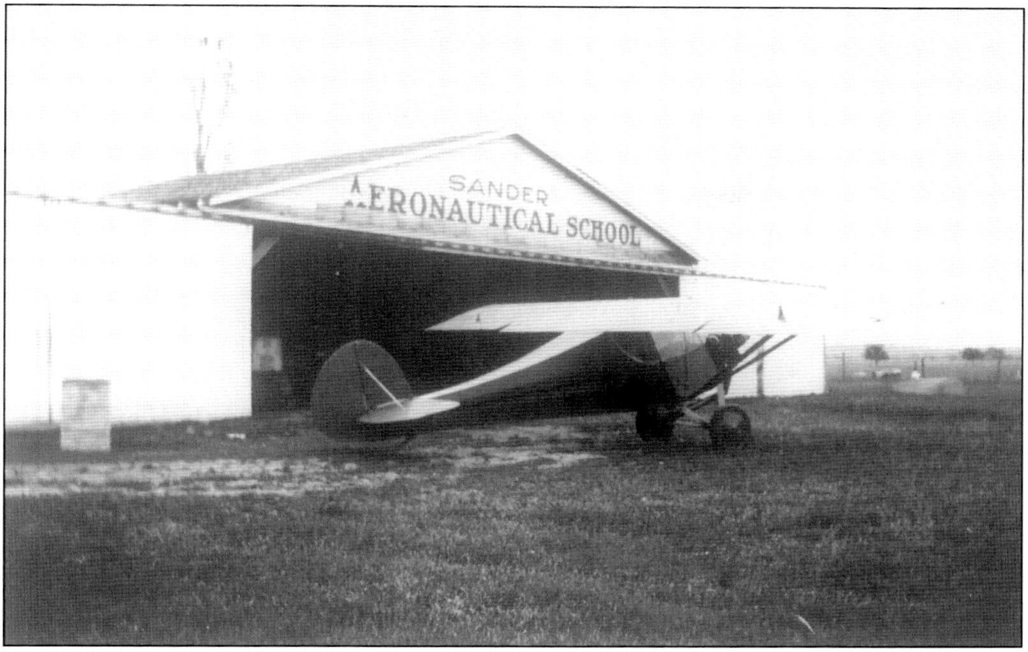

Harold Sander was a friend of Pacific Air Transport pilot Art Starbuck. Starbuck located an old Jenny aircraft and Sander came up with the cash to buy it, then learned how to fly from his friend. In 1927, Sander opened the Sander Aeronautical School at Newell barber field and began giving flying lessons for $25 per hour. (SOHS No. 12729.)

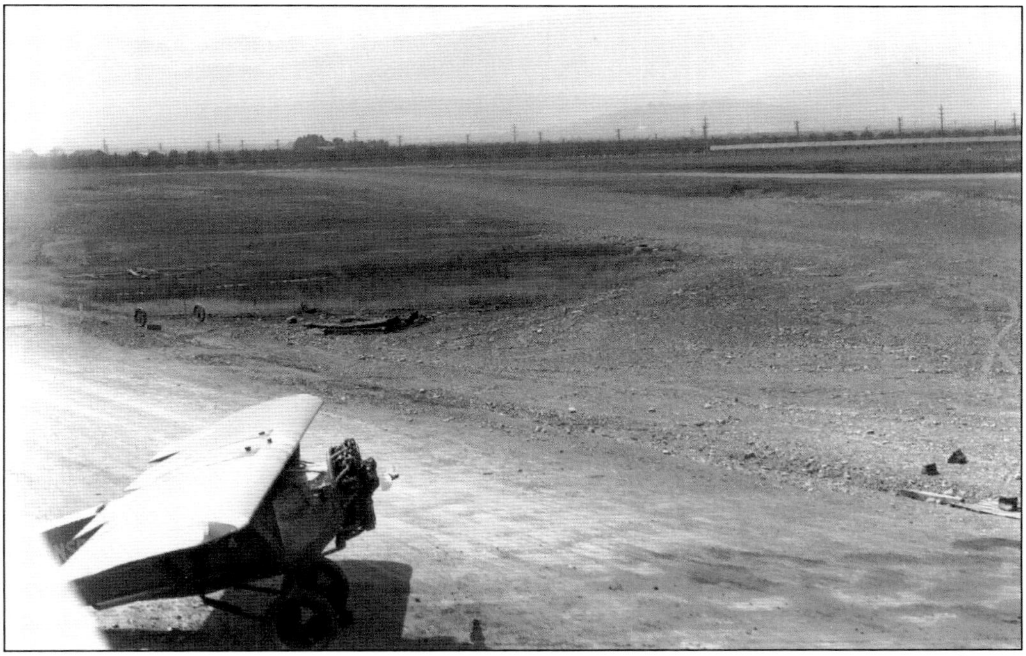

The primitive state of Newell Barber Field is seen in this image taken from the roof of Pacific Air Transport's hangar. A Ryan M-1 mail plane sits in the foreground; the Jackson County Fairgrounds are not visible but to the right. (SOHS No. 19206.)

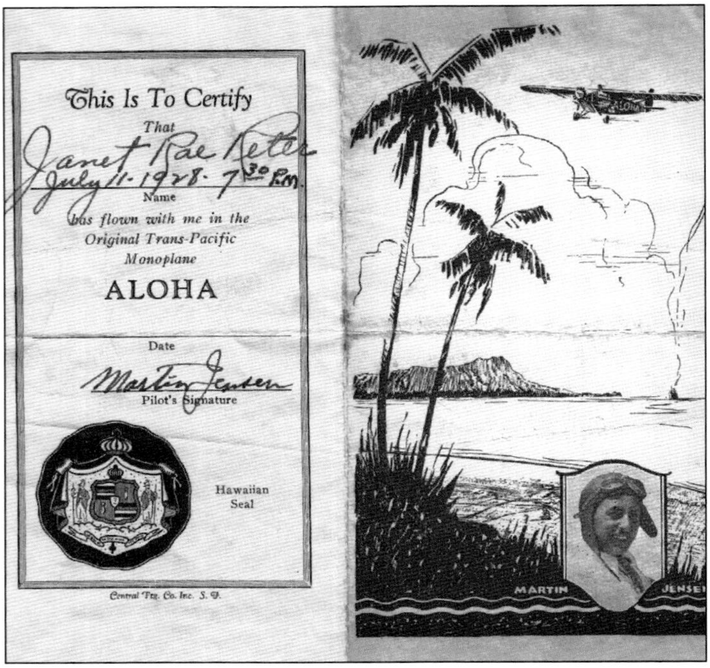

Martin Jensen was one of the best known aviators of the 1920s. In 1927, he placed second in the tragedy-marred Dole Transpacific Race between Oakland, California, and Honolulu, Hawaii. Ten people were killed, and only two planes completed the flight. The following year, he took his bright yellow Breese Monoplane, the *Aloha*, on a nationwide tour. During his stop in Medford, he raised money by offering rides, which included a souvenir certificate. One of his passengers was Janet Reter, who saved her certificate. (SOHS Acc. No. 1999.135.)

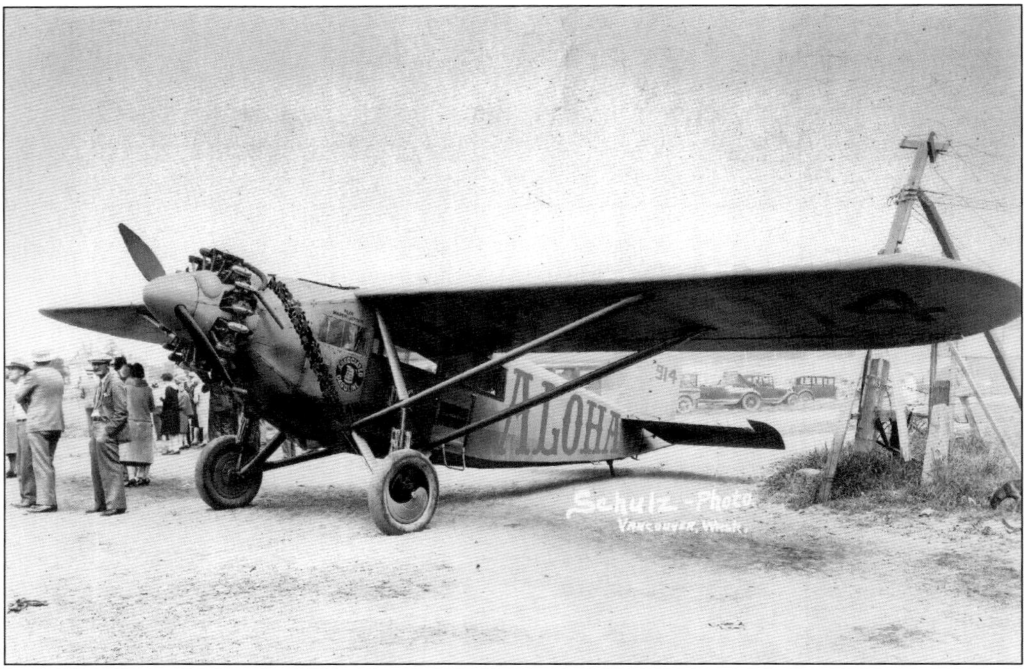

A month after the Dole Transpacific Race, Jensen was again in the news. He was flying a celebrity to New York City when his plane went down in a remote arroyo in Arizona. Though neither was injured, Jensen's passenger, Leo, the MGM lion, gave him a "thoroughly disgusted look." Fortunately the cage in the specially modified Ryan remained intact. The *Aloha* fared much better on its nationwide tour. (Pearson Air Museum.)

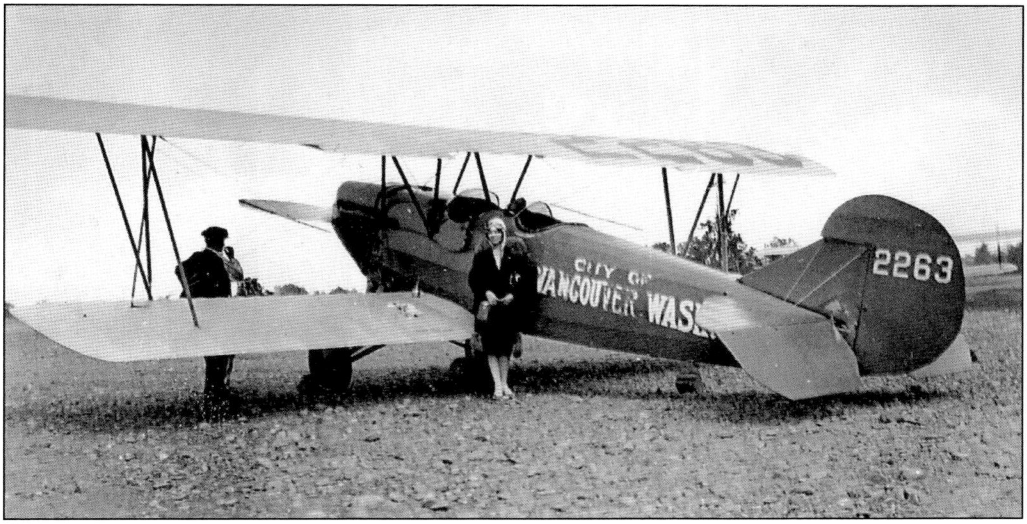

As part of the National Air Races for 1927, a sanctioned race was organized from San Francisco, California, to Spokane, Washington, with a stop in Medford. Pacific Air Transport pilot Vern Bookwalter, who had delivered the first airmail to Medford the year before, entered his Travel Air in the competition, which was sponsored by the chamber of commerce of Vancouver, Washington. (Pearson Air Museum.)

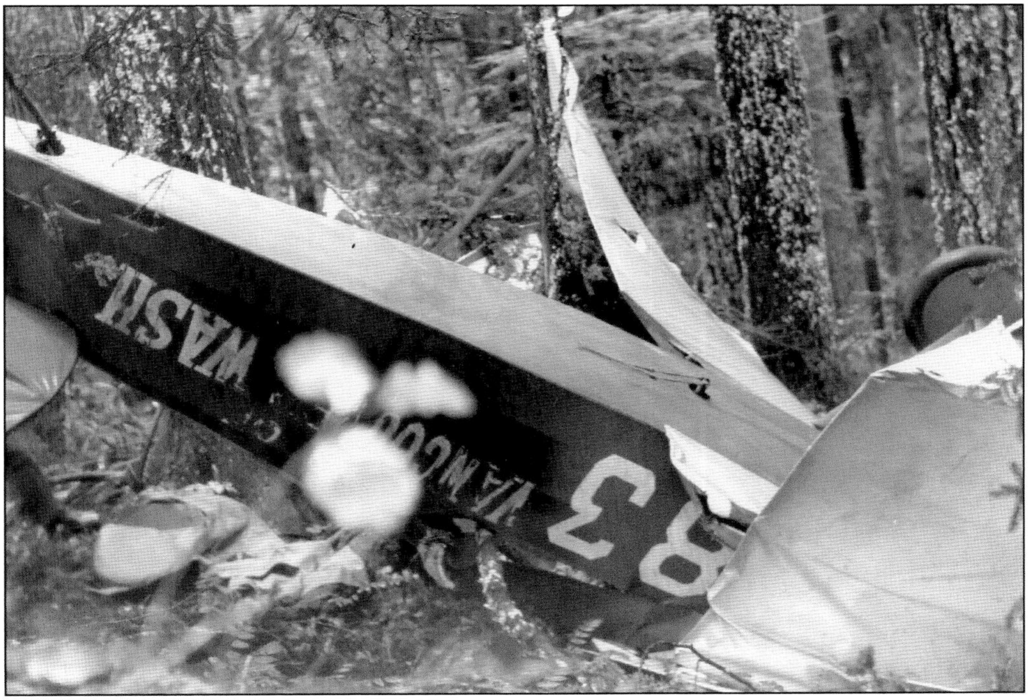

Bookwalter and his plane, the *City of Vancouver,* were well in the lead of the 1927 National Air Race when the competing planes landed at Newell Barber Field. After leaving Medford, however, heavy fog hung south of Eugene, and while trying to fly below it, Bookwalter's engine failed. The veteran Pacific Air Transport pilot escaped without injury, but his plane was a total wreck. (Pearson Air Museum.)

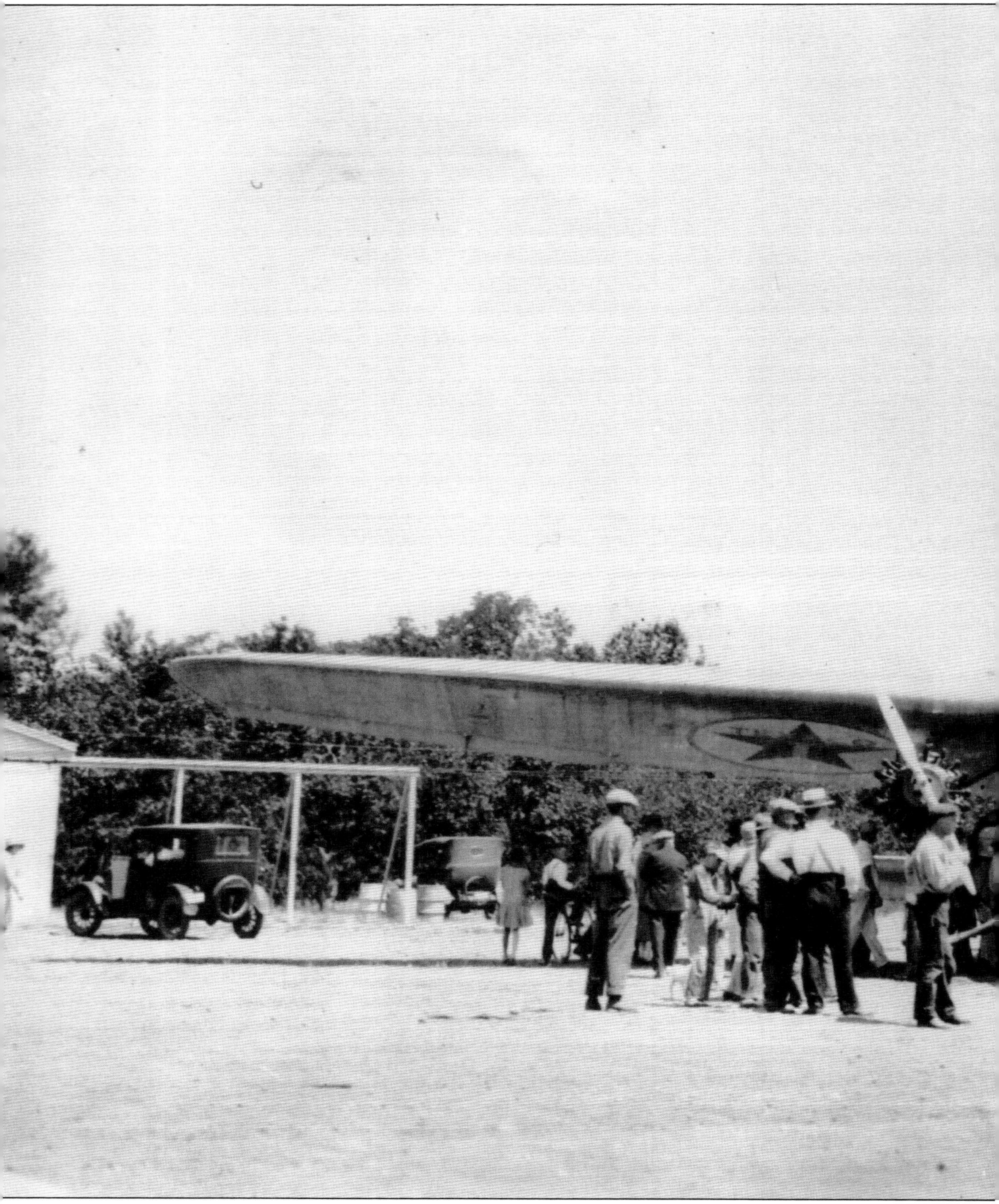

The Texas Oil Company, now known as Texaco, introduced its line of aviation products in 1927. To promote these products, the company acquired a custom-built Ford trimotor named *Texaco 1*. In 1928, famed pilot Frank Hawkes embarked on a promotional tour around the country, visiting

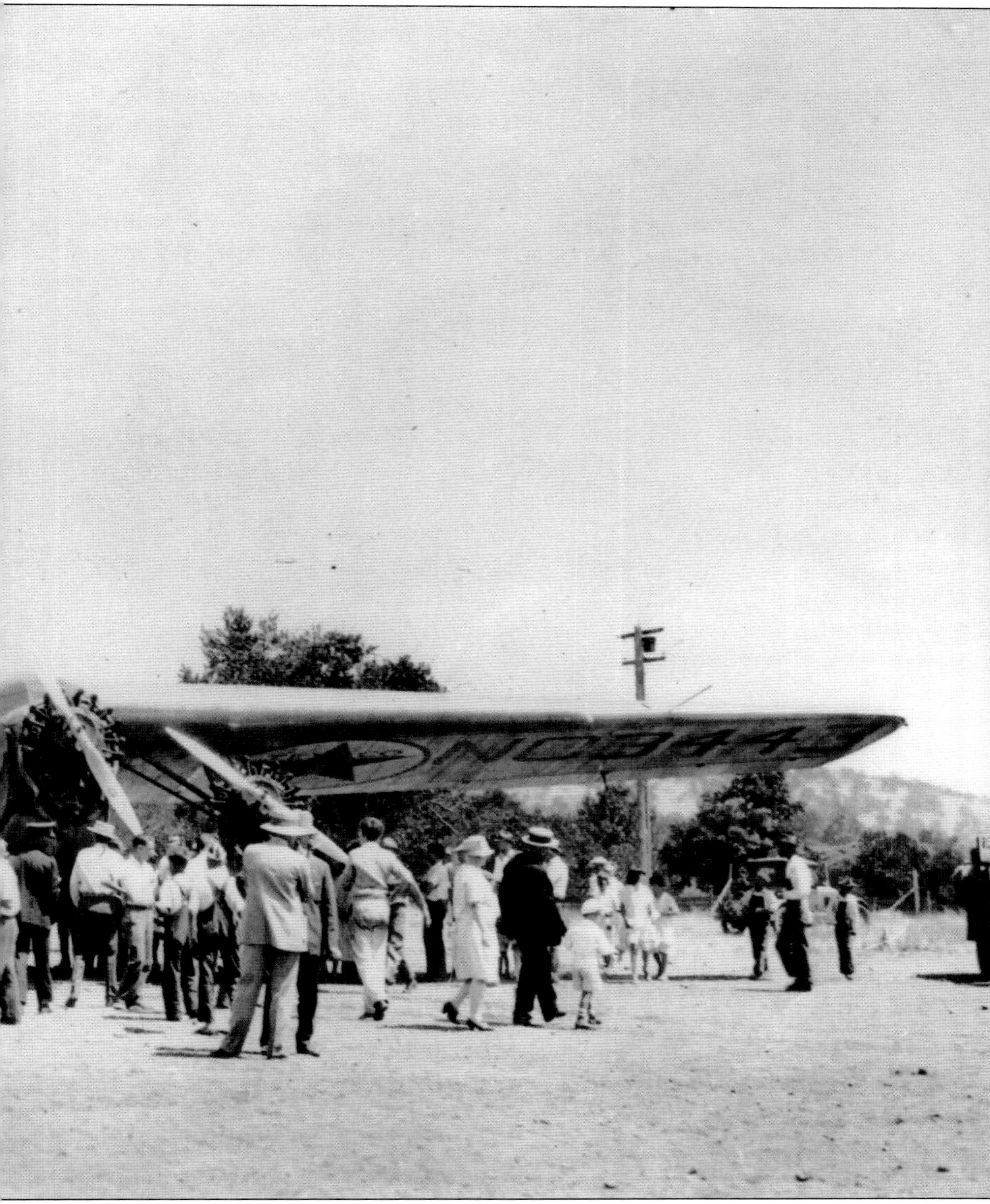

175 cities and providing flights for 7,200 passengers. Among his many stops that summer was Medford's Newell Barber Field. (SOHS No. 9199.)

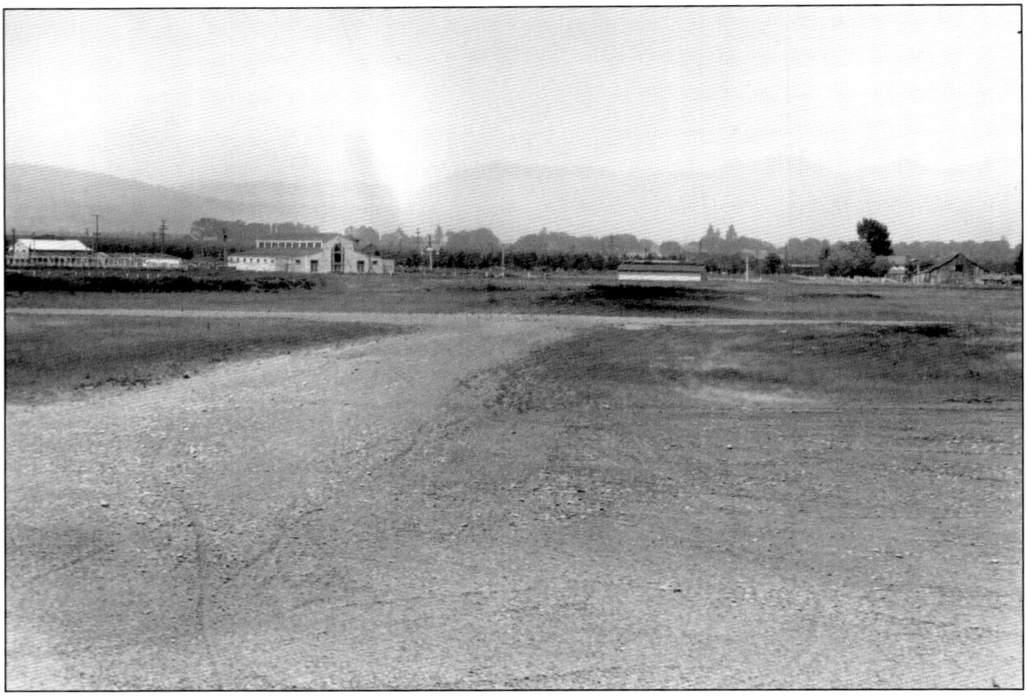

Newell Barber Field was adjacent to the Jackson County Fairgrounds, just south of downtown off the Pacific Highway. This view, looking west from the Pacific Air Transport hangar, shows some of the fairground's buildings around 1926. (SOHS No. 19210.)

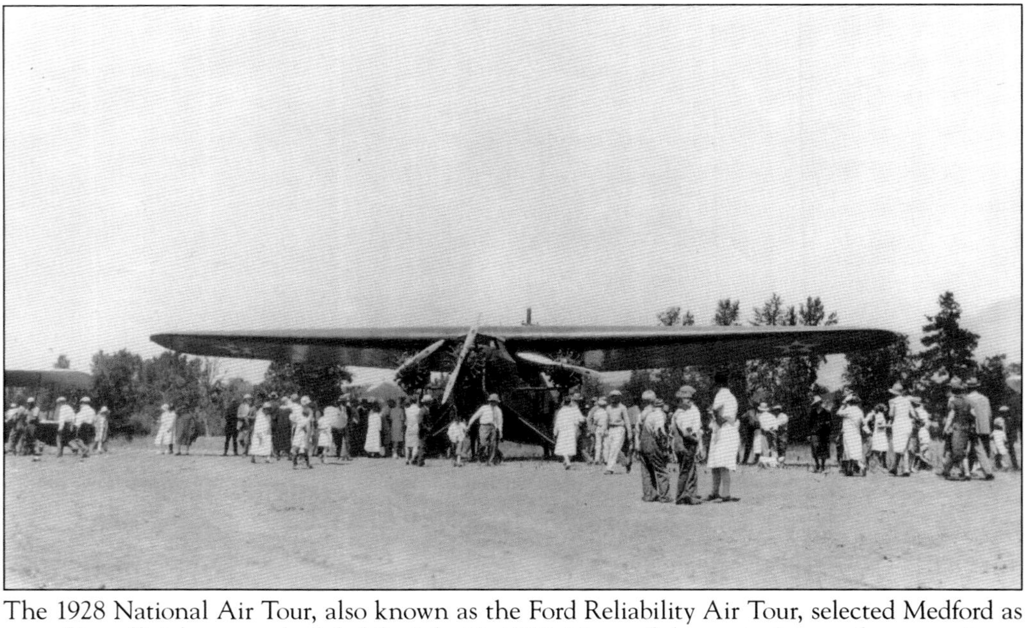

The 1928 National Air Tour, also known as the Ford Reliability Air Tour, selected Medford as one of its official stops, scheduled for July 16, 1928. Many of the nation's most famous aviators and some of the newest aircraft participated in this tour; much of the attention was focused on the "giant" eight-passenger Ford transport plane. (SOHS No. 9194.)

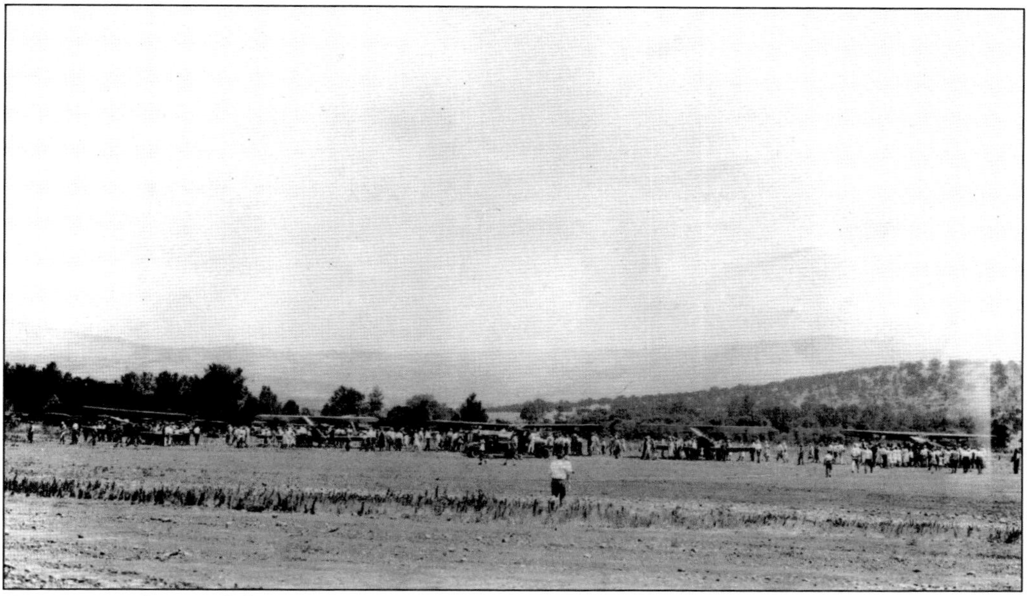

Twenty-six planes of the 1928 National Air Tour visited Newell Barber Field, which was the smallest field on the tour. While the local crowds delighted in seeing the myriad of aircraft and famous aviators, there was some grumbling among pilots of larger aircraft that Newell Barber Field was not a particularly good airstrip. (SOHS No. 9202.)

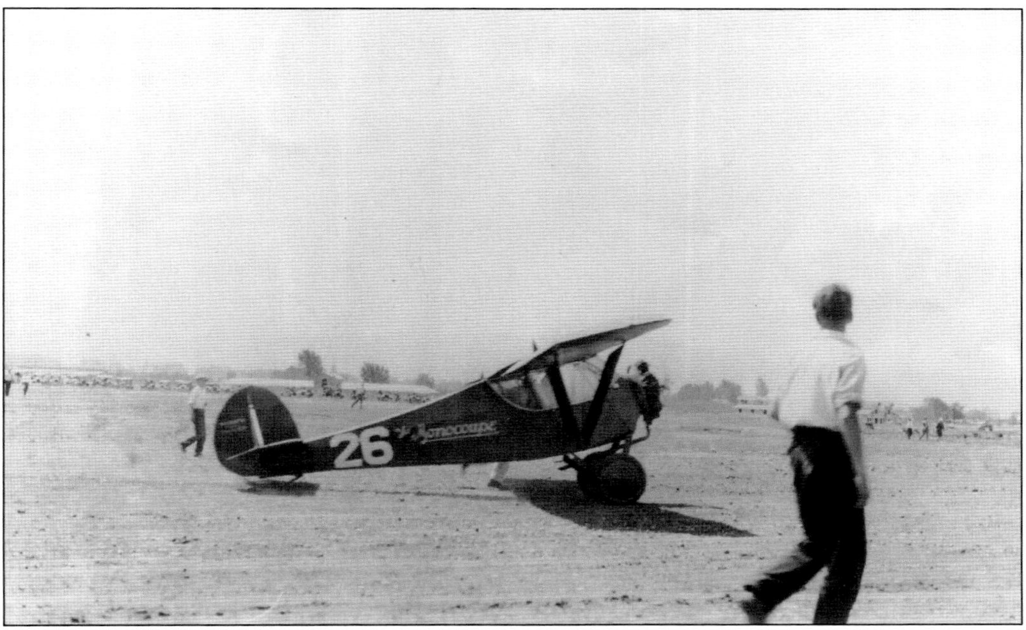

One of the many famous pilots participating in the 1928 National Air Tour, and the tour's only female pilot, was Phoebe Omlie. Omlie, the nation's first licensed female transport pilot, flew plane No. 26, a Monocoupe 70. Omlie later served as a pilot flying female speakers supporting Franklin Roosevelt during the 1932 presidential campaigns. (SOHS No. 9195.)

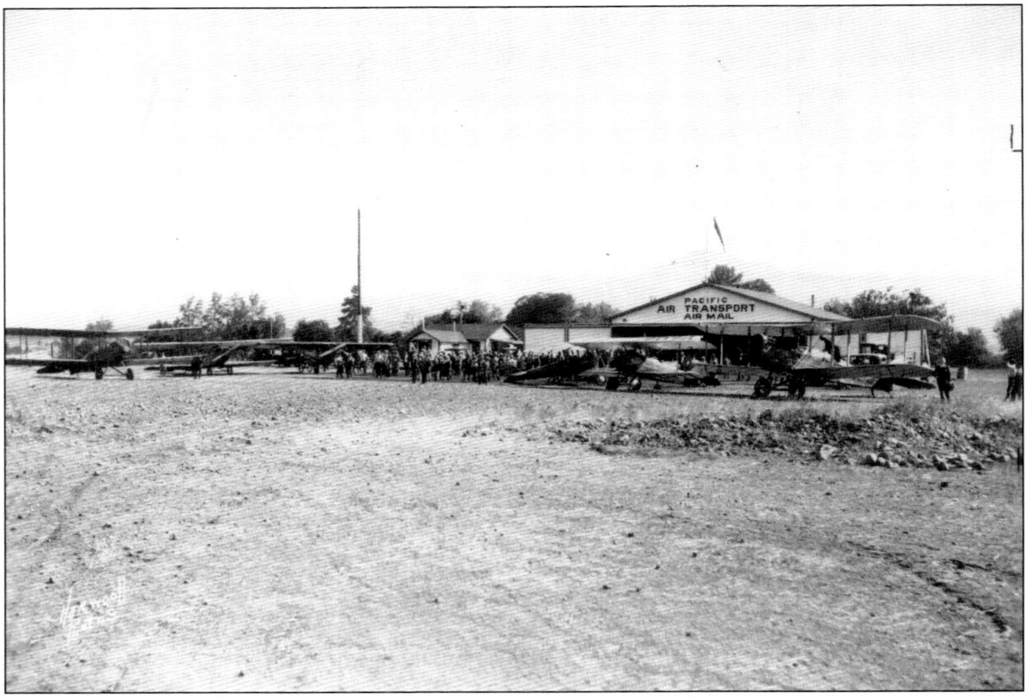

The Pacific Air Transport hangar and office was the only developed portion of Newell Barber Field, and the participants in the National Air Tour gathered around that portion of the field. (SOHS No. 928.)

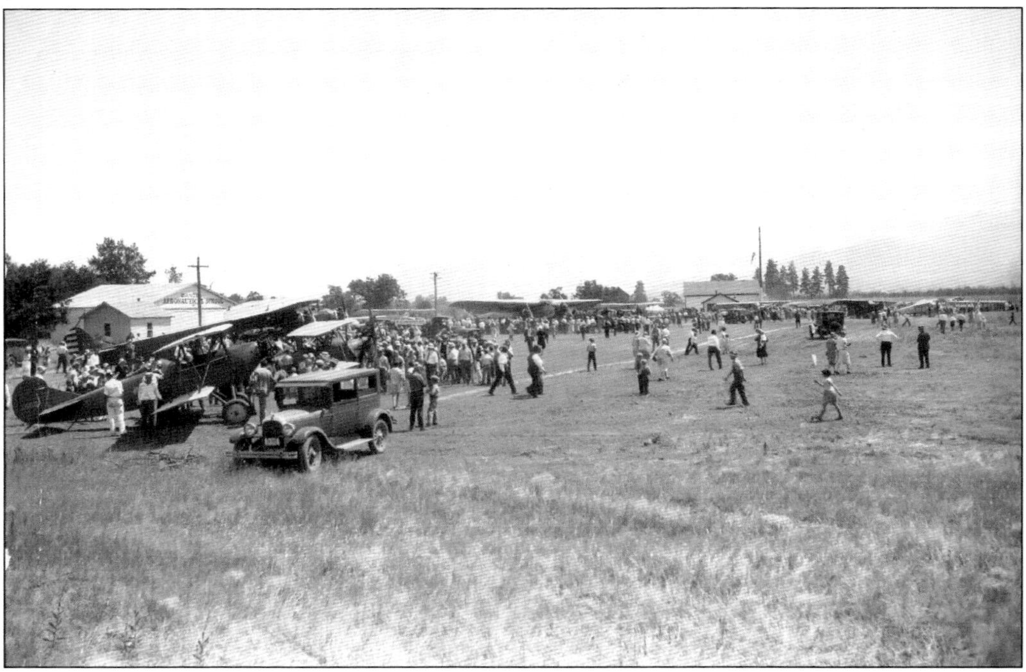

Spectators gather to examine the aircraft of the National Air Tour. The developed area between the Sander Aeronautical School and Pacific Air Transport was a beehive of activity. (SOHS.)

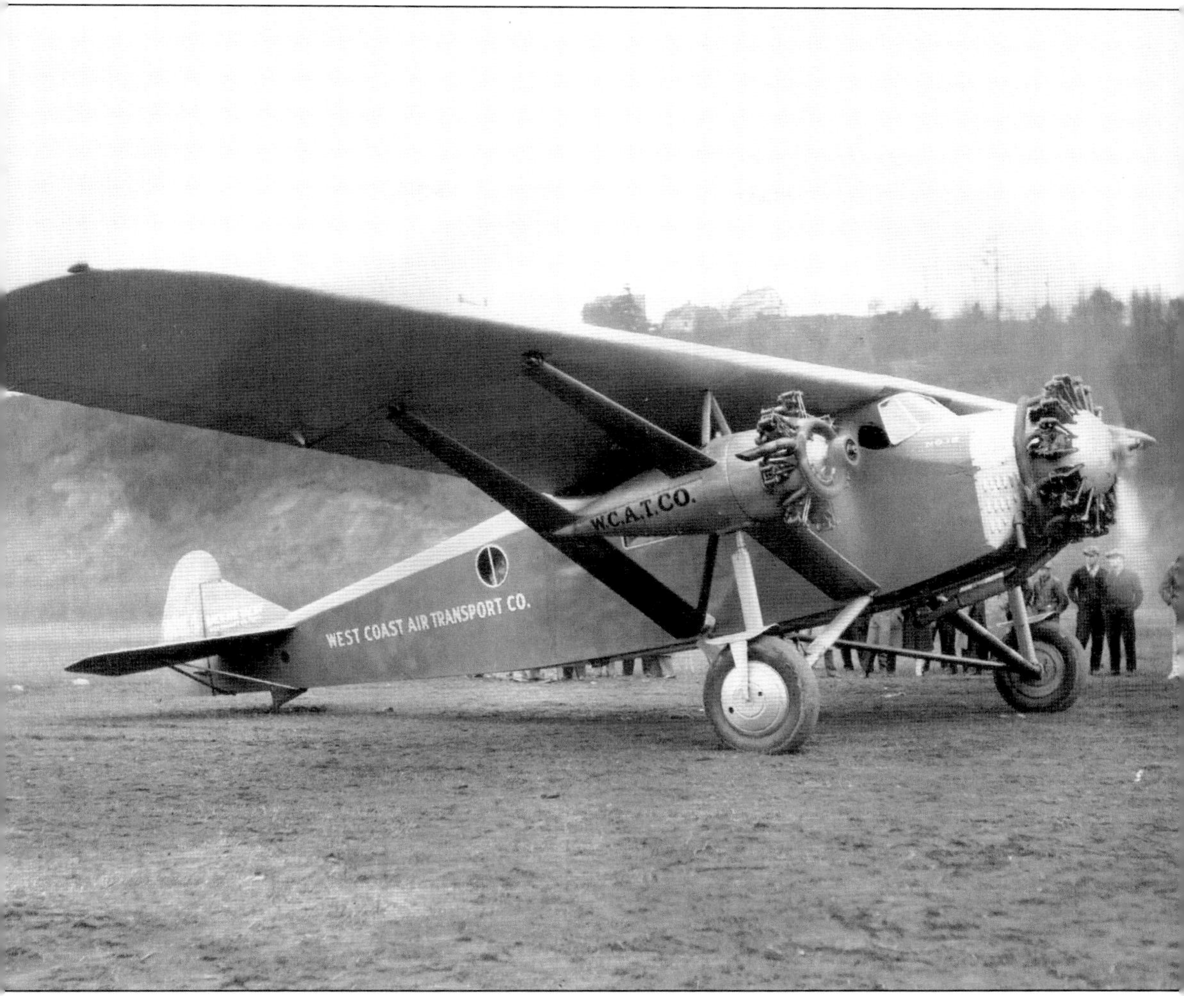

A new commercial carrier came to Newell Barber Field in 1928. West Coast Air Transport, a subsidiary of the Pickwick Stage Lines, began carrying passengers in a trimotor Bach Air Yacht. West Coast Air Transport abandoned Medford for Eugene when it expanded its fleet to include the Fokker trimotor aircraft because Medford's field was too small and undeveloped. This was the first in a series of events that resulted in the creation of the new Medford airport. This image shows a West Coast Air Yacht. (Pearson Air Museum.)

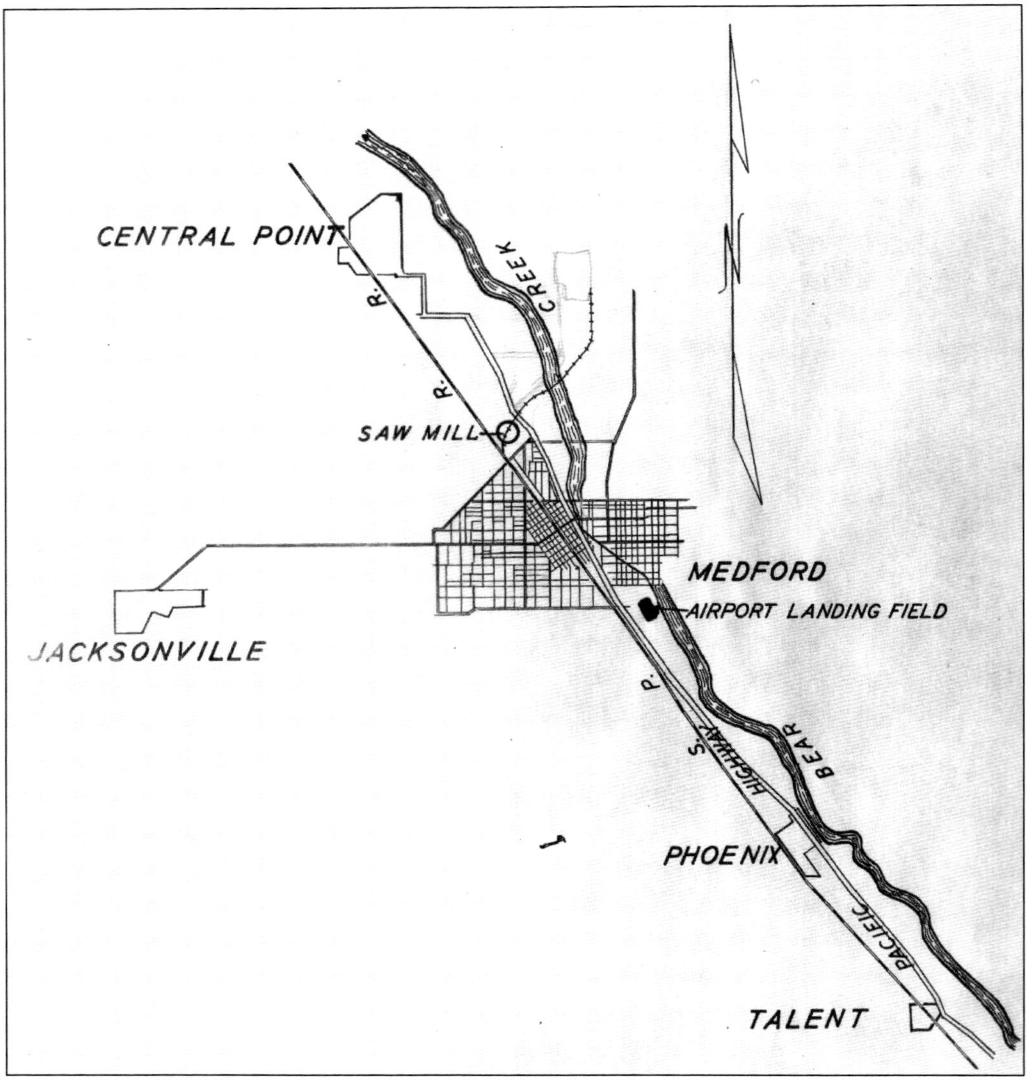

This map pinpoints the location of Newell Barber Field in relation to Medford and other nearby communities. This map illustrates how the field, wedged in between Bear Creek and the fairgrounds on the Pacific Highway, made expansion impossible and forced relocation. It appears as if someone has penciled in the approximate location of the new airport north of town. (Rogue Valley International-Medford Airport.)

Three

Pacific Air Transport

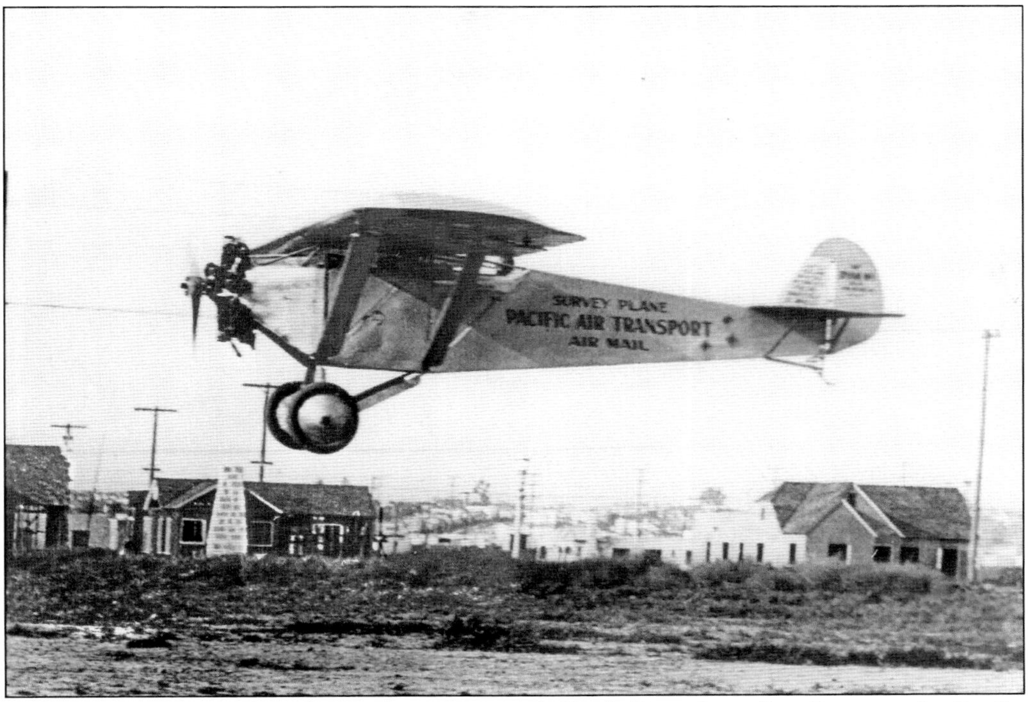

Vern Gorst, founder of Pacific Air Transport, and Claude Ryan leave San Diego in the Pacific Air Transport survey plane, a Ryan M-1, in 1926. Written on the plane's tail are all of the cities and towns that will be served by the new airmail service along Contract Airmail Route 8. (SOHS No. 959.)

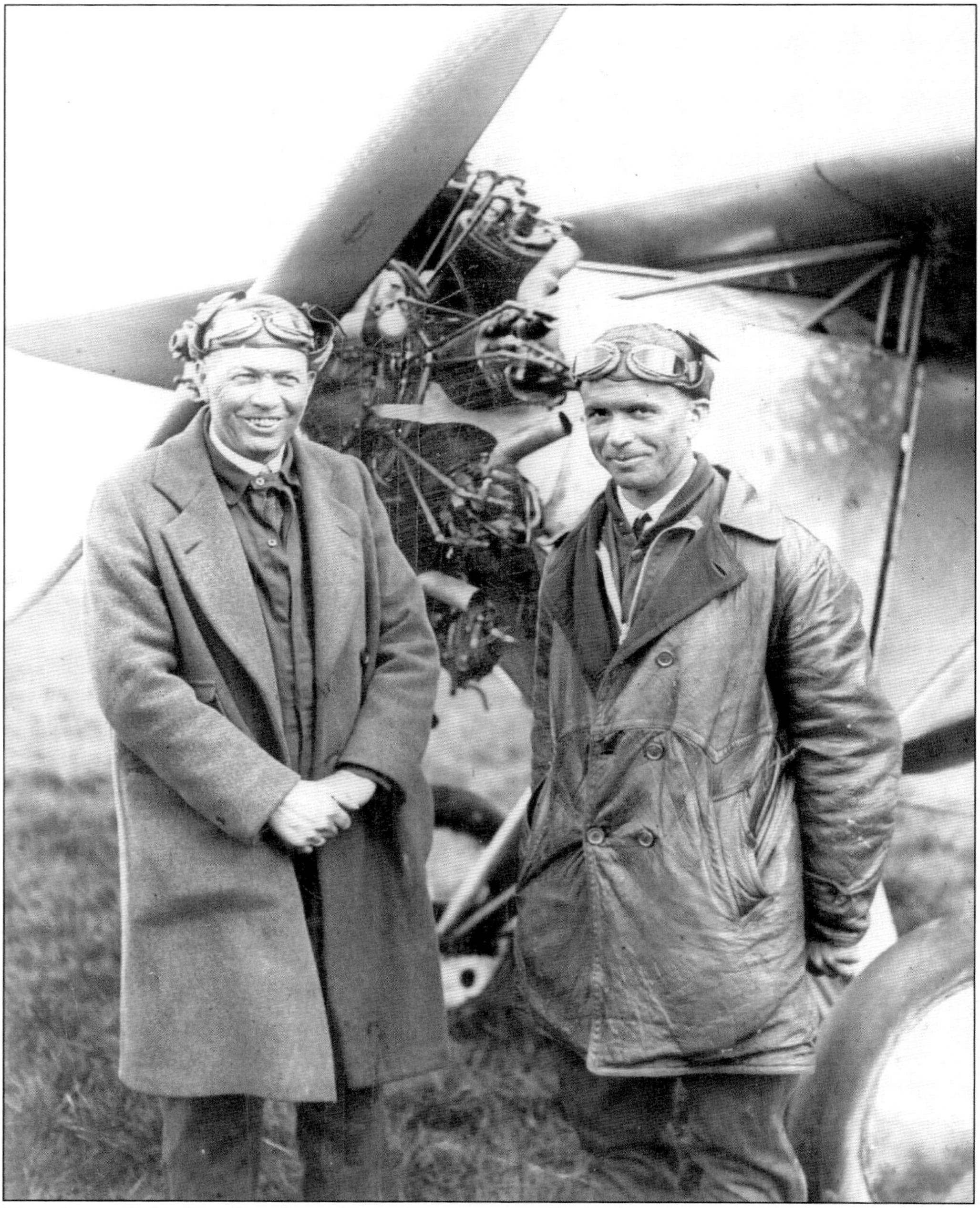

Vern Gorst (left), founder of Pacific Air Transport, and Claude Ryan (right) of the Ryan Aircraft Company, made an aerial survey of the recently awarded Contract Air Mail Route 8 (CAM) in early 1926. Gorst selected the Ryan M-1 monoplane as a major component of his fleet. (SOHS No. 23570.)

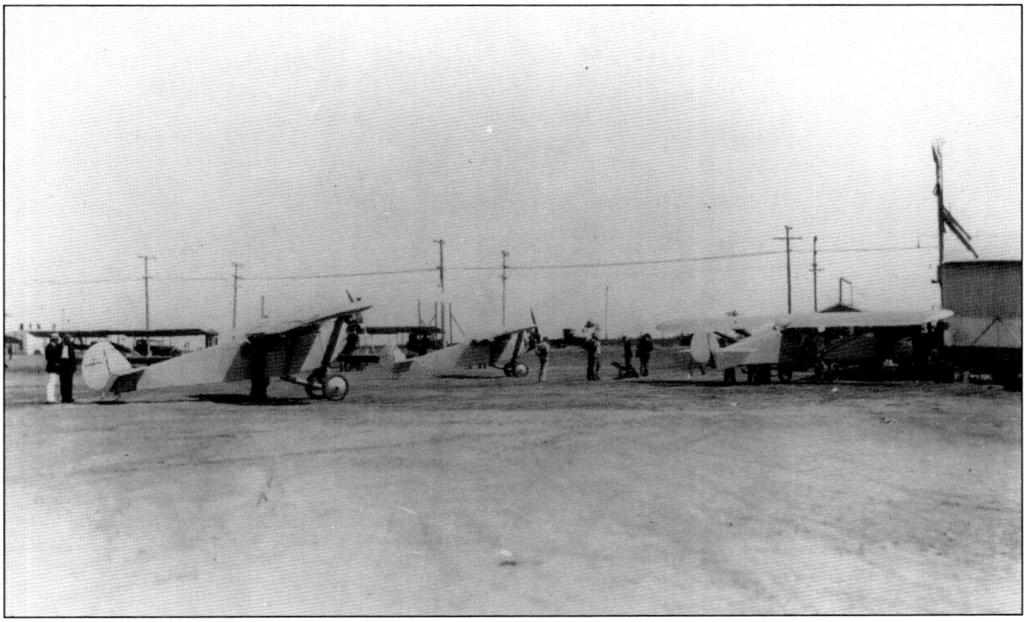

Three M-1 mail planes are lined up at the Ryan Aircraft plant in San Diego. Vern Gorst, founder of Pacific Air Transport, contracted for 10 Ryan mail planes to service his West Coast airmail route, Contract Airmail Route 8. (SOHS No. 20644.)

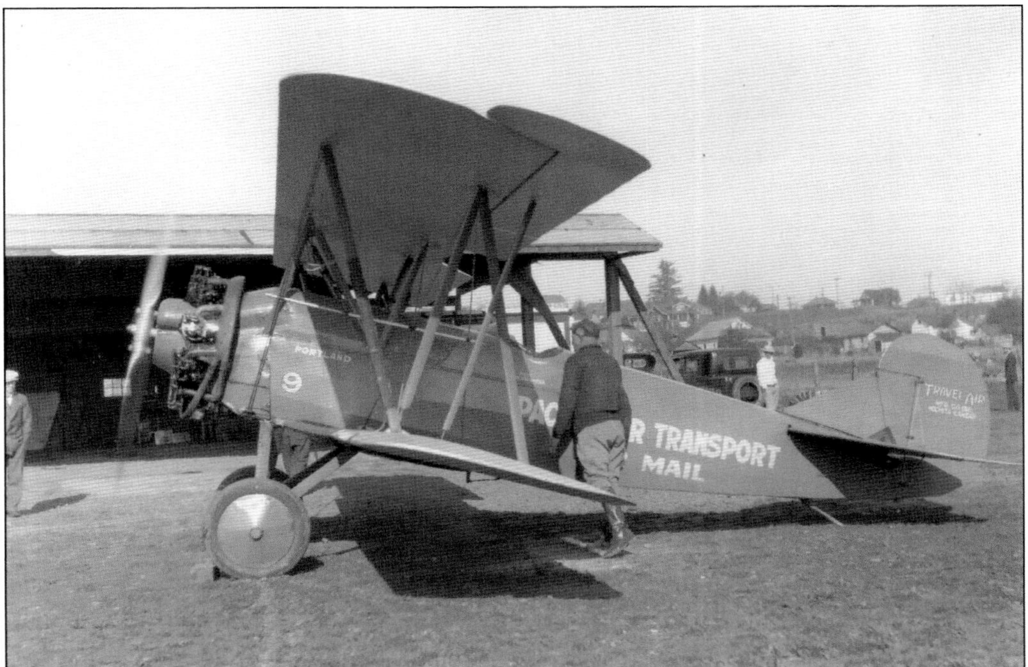

Veteran Pacific Air Transport pilot Vern Bookwalter was based at Vancouver's Pearson Field and flew the regular run between Pearson and Medford. As demand for passenger space grew and aircraft design improved, Bookwalter's Travelaire biplane was replaced by a Fokker Universal with cabin space for four people. The pilot, of course, remained out in the open. (Pearson Air Museum.)

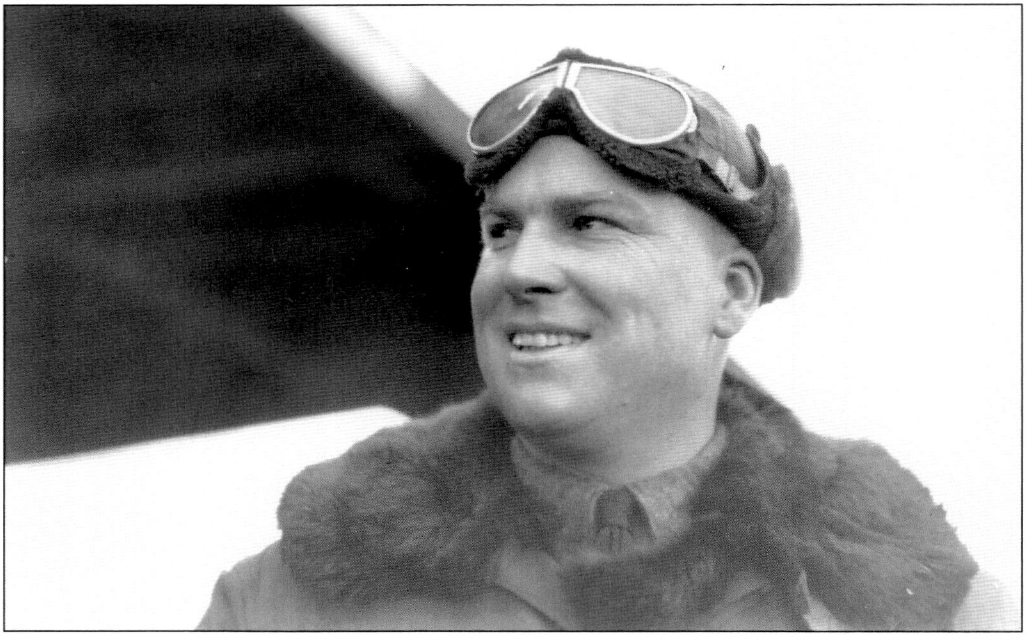

Oregon native Vernon Bookwalter (1898–1967) was one of the original Pacific Air Transport pilots hired by Vern Gorst. On September 15, 1926, he arrived in Medford with the first delivery of airmail and crashed without injury near Eugene during the San Francisco to Spokane air race in 1927. He eventually moved on to Alaska, where he became an accomplished bush pilot. (Pearson Air Museum.)

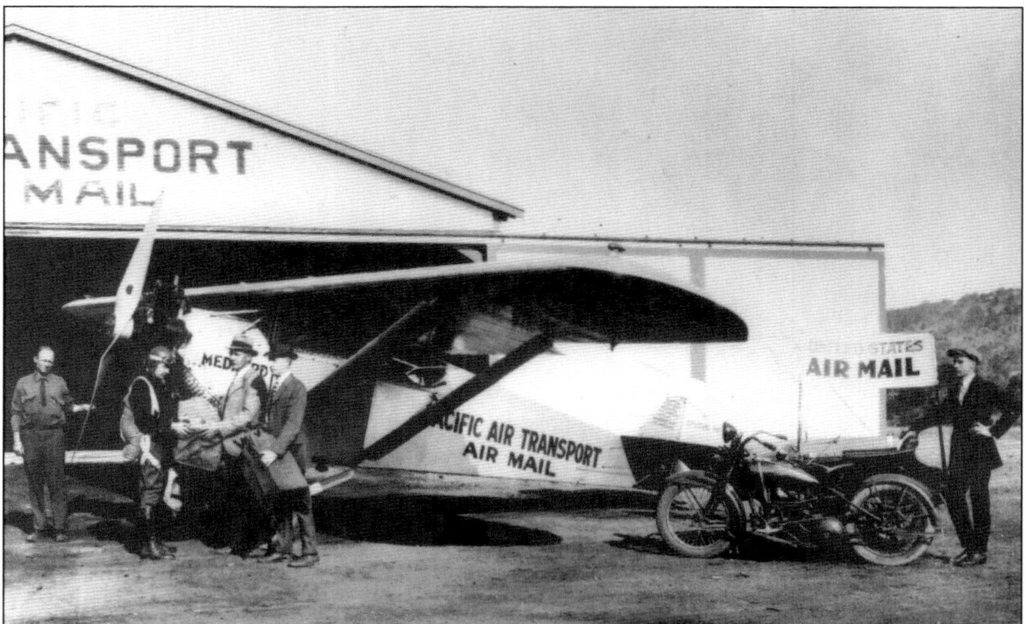

Pilot Pat Patterson is handed the first mail sacks by postal director William Warner and Medford postmaster Roland Beach on the morning of September 15, 1926. The Pacific Air Transport chief mechanic stands by the propeller and Ernie Scott is by the motorcycle. (SOHS No. 004.)

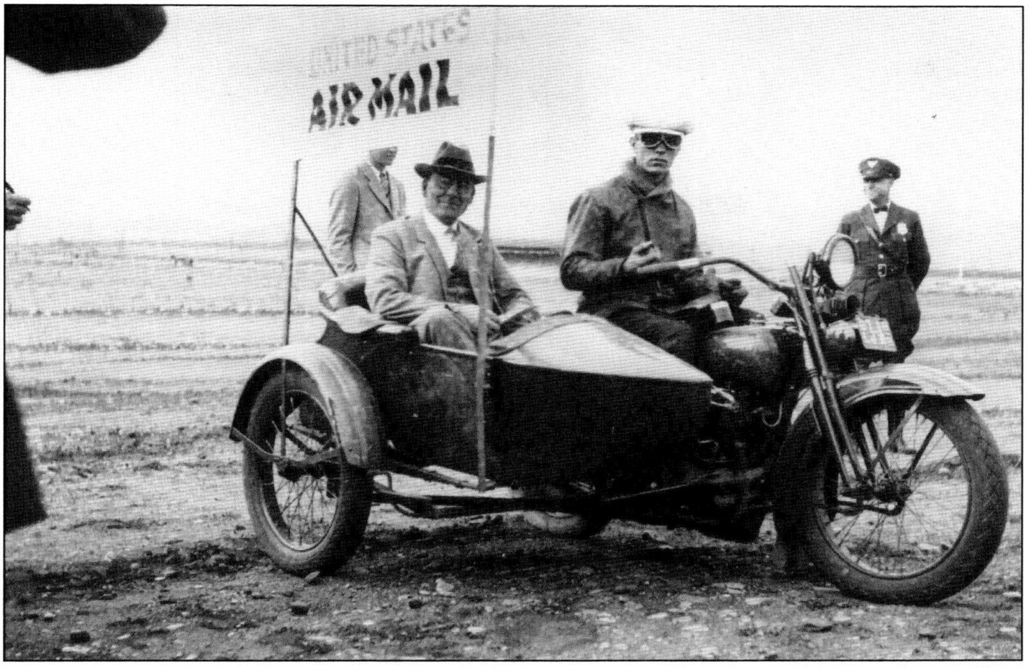

The first ceremonial sack of airmail was whisked out to Newell Barber Field from the Medford post office in a 1924 Harley Davidson motorcycle and sidecar. (SOHS No. 8616.)

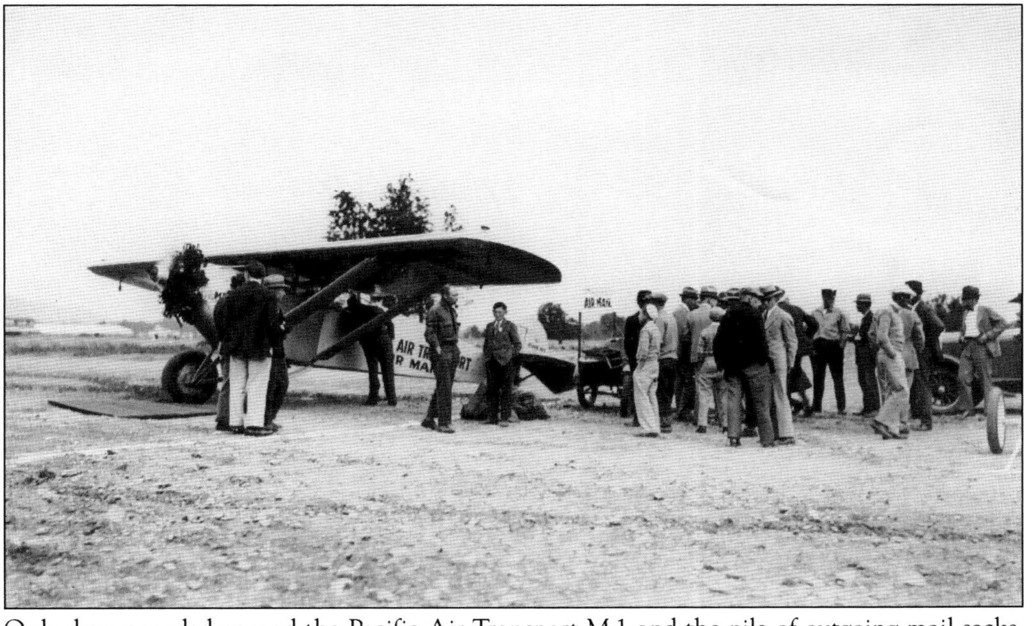

Onlookers crowded around the Pacific Air Transport M-1 and the pile of outgoing mail sacks. The subsidy for airmail was determined by weight and the mailbags were constructed of heavy canvas and secured with a large, weighty brass lock. The inherent dangers of flying the mail became apparent not long after this flight when pilot Pat Patterson's plane crashed in the Siskiyou Mountains and he was killed. (SOHS No. 8617.)

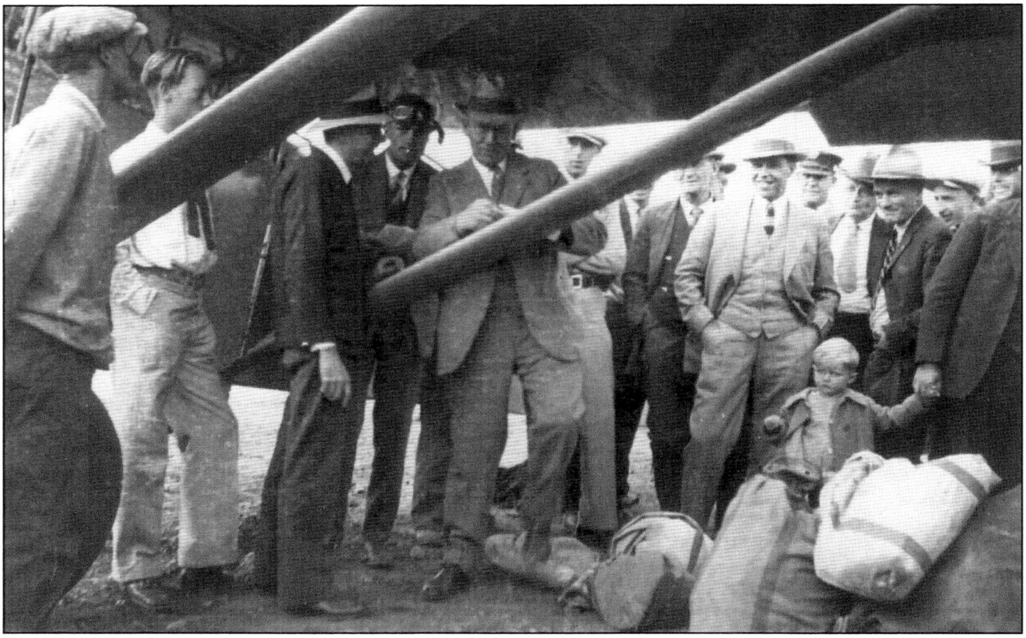

A crowd gathers under the wing of Pat Patterson's Ryan mail plane before the canvas sacks of mail are loaded. The pilot, in the background, enjoys a cigarette before his return trip to Oakland. (SOHS No. 8615.)

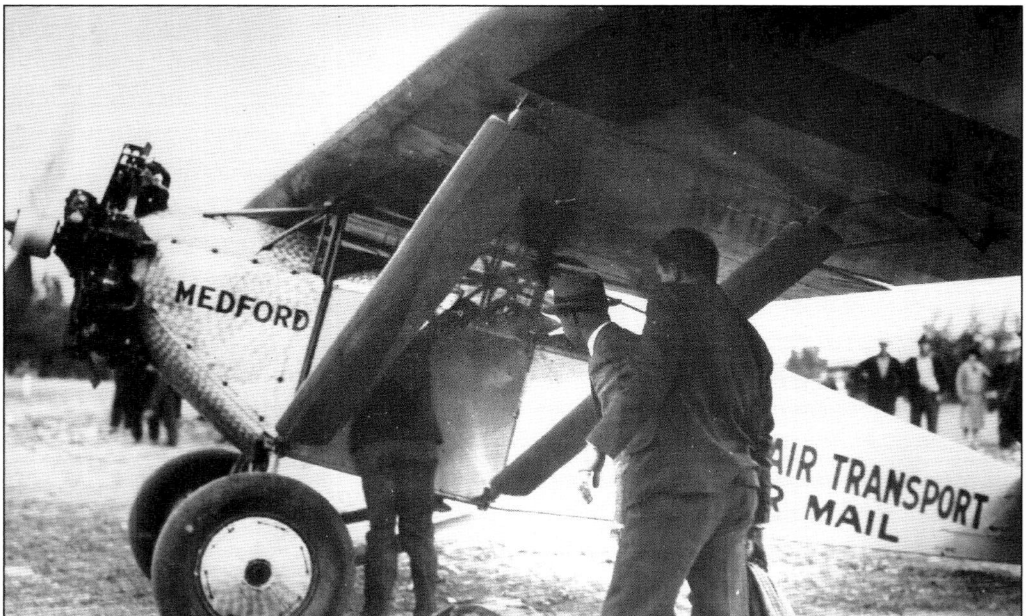

The selection of Newell Barber Field in 1926 as Oregon's only airmail stop generated an enormous amount of excitement. Pacific Air Transport pilot Vern Bookwalter arrived in the morning with the first delivery, but it was Pat Patterson, in a Ryan M-1 mail plane named for the city of Medford, who garnered most of the attention when he carried the first outgoing mail. (SOHS No. 8614.)

The coming of regularly scheduled airmail service generated a great deal of excitement in southern Oregon, especially since Newell Barber Field was the only airmail stop in the state. A popular souvenir of the occasion was a first flight letter, sponsored by the American Legion and the Crater Lake Council of the Boy Scouts, bearing special commemorative stamps and carried to Medford on the first delivery on September 15, 1926. (SOHS Manuscript.)

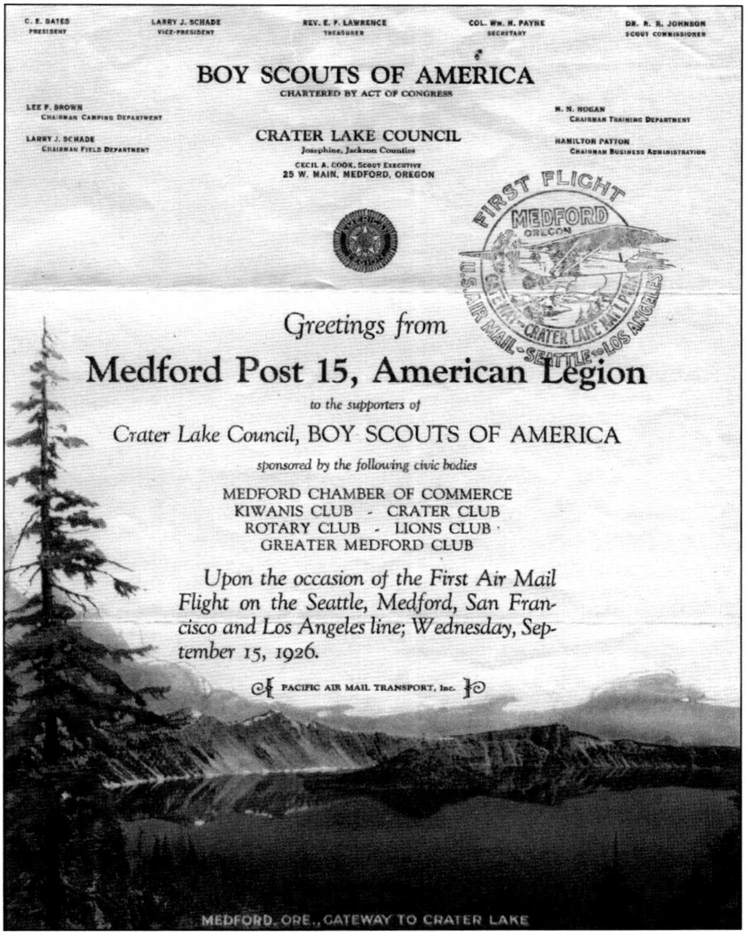

C. E. BATES, PRESIDENT · LARRY J. SCHADE, VICE-PRESIDENT · REV. E. P. LAWRENCE, TREASURER · COL. WM. H. PAYNE, SECRETARY · DR. R. R. JOHNSON, SCOUT COMMISSIONER

BOY SCOUTS OF AMERICA

CHARTERED BY ACT OF CONGRESS

LEE P. BROWN, CHAIRMAN CAMPING DEPARTMENT

LARRY J. SCHADE, CHAIRMAN FIELD DEPARTMENT

CRATER LAKE COUNCIL

Josephine, Jackson Counties

CECIL A. COOK, SCOUT EXECUTIVE

25 W. MAIN, MEDFORD, OREGON

H. N. HOGAN, CHAIRMAN TRAINING DEPARTMENT

HAMILTON PATTON, CHAIRMAN BUSINESS ADMINISTRATION

FIRST FLIGHT MEDFORD OREGON U.S. AIR MAIL · SEATTLE · LOS ANGELES · GATEWAY · CRATER LAKE NAT'L PARK

Greetings from

Medford Post 15, American Legion

to the supporters of

Crater Lake Council, BOY SCOUTS OF AMERICA

sponsored by the following civic bodies

MEDFORD CHAMBER OF COMMERCE

KIWANIS CLUB - CRATER CLUB

ROTARY CLUB - LIONS CLUB

GREATER MEDFORD CLUB

Upon the occasion of the First Air Mail Flight on the Seattle, Medford, San Francisco and Los Angeles line; Wednesday, September 15, 1926.

PACIFIC AIR MAIL TRANSPORT, Inc.

MEDFORD, ORE., GATEWAY TO CRATER LAKE

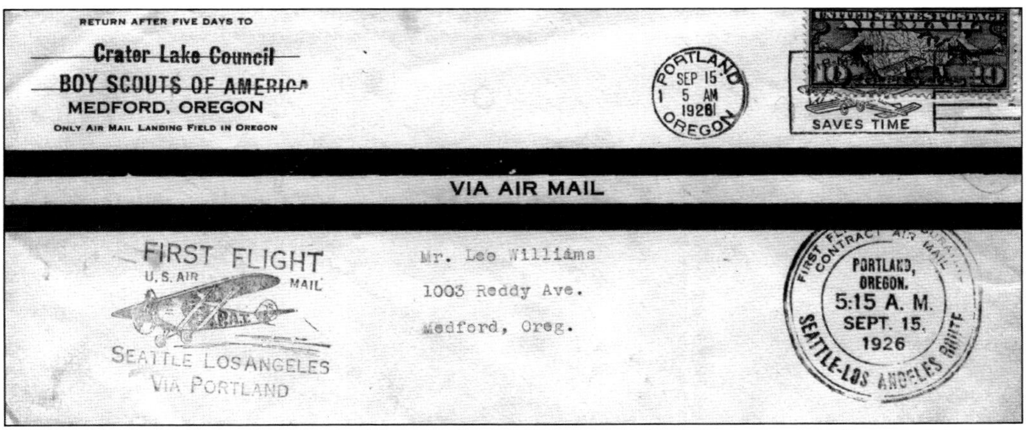

Shown here is a specially cancelled first flight envelope from the first airmail flight in to Medford. This envelope contained the letter in the previous picture. (SOHS Manuscript.)

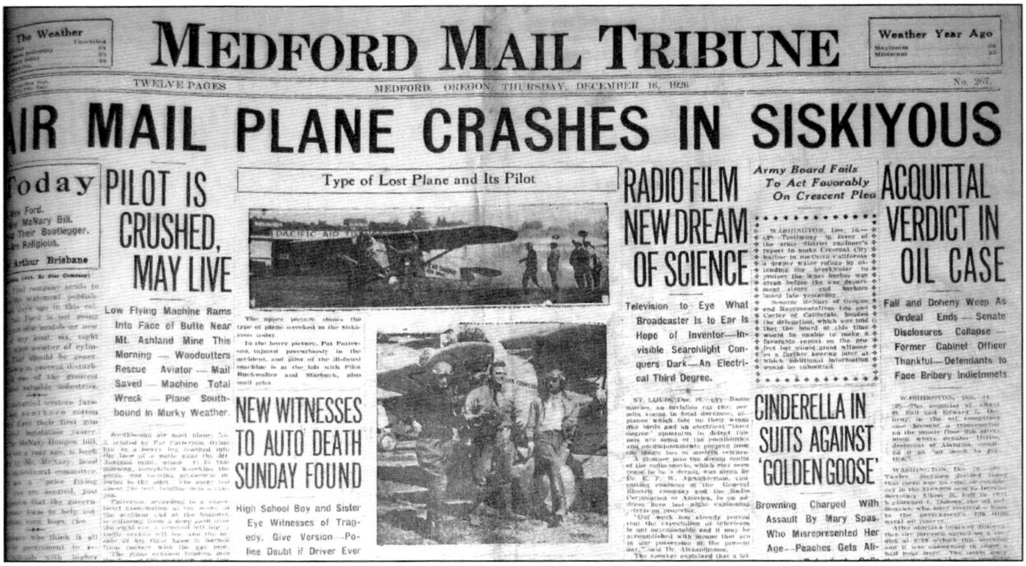

MEDFORD MAIL TRIBUNE

TWELVE PAGES — MEDFORD, OREGON, THURSDAY, DECEMBER 16, 1926 — No. 267

AIR MAIL PLANE CRASHES IN SISKIYOUS

Today — Arthur Brisbane

PILOT IS CRUSHED, MAY LIVE

Low Flying Machine Rams Into Face of Butte Near Mt. Ashland Mine This Morning — Woodcutters Rescue Aviator — Mail Saved — Machine Total Wreck — Plane Southbound in Murky Weather.

Type of Lost Plane and Its Pilot

NEW WITNESSES TO AUTO DEATH SUNDAY FOUND

High School Boy and Sister Eye Witnesses of Tragedy, Give Version—Police Doubt if Driver Ever

RADIO FILM NEW DREAM OF SCIENCE

Television to Eye What Broadcaster Is to Ear Is Hope of Inventor—Invisible Searchlight Conquers Dark—An Electrical Third Degree.

Army Board Fails To Act Favorably On Crescent Plea

CINDERELLA IN SUITS AGAINST 'GOLDEN GOOSE'

Browning Charged With Assault By Mary Spas. Who Misrepresented Her Age—Peaches Gets Ali-

ACQUITTAL VERDICT IN OIL CASE

Fall and Doheny Weep As Ordeal Ends — Senate Disclosures Collapse — Former Cabinet Officer Thankful—Defendants to Face Bribery Indictments

The dangers inherent to airmail operations were brought home tragically a mere two months after the inaugural flight. Pat Patterson, who had carried the first bags of mail from Medford on September 15, was fatally injured when his southbound mail flight crashed in the Siskiyou Mountains south of Ashland. (*Medford Mail Tribune*, December 16, 1926.)

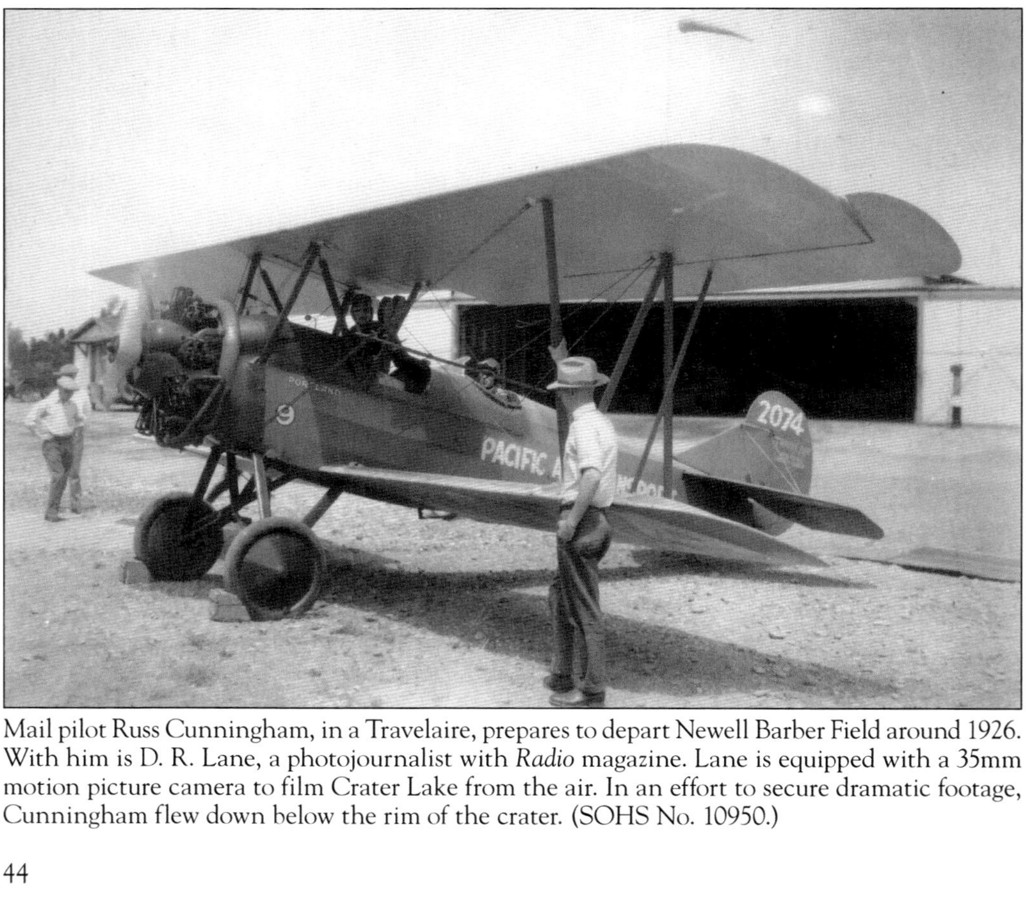

Mail pilot Russ Cunningham, in a Travelaire, prepares to depart Newell Barber Field around 1926. With him is D. R. Lane, a photojournalist with *Radio* magazine. Lane is equipped with a 35mm motion picture camera to film Crater Lake from the air. In an effort to secure dramatic footage, Cunningham flew down below the rim of the crater. (SOHS No. 10950.)

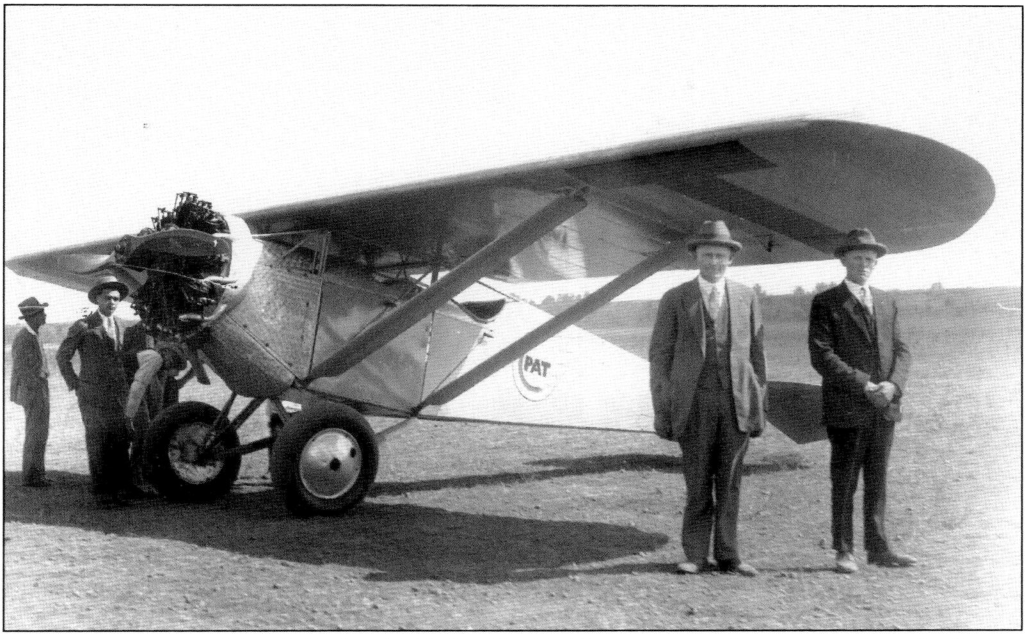

During its tenure at Newell Barber Field, Pacific Air Transport employed a number of different types of aircraft. Most common was the Ryan M-1, but five of the first six crashed during the airline's first year of operations. Some of the newer Ryans had a squared-off tail in place of the original rounded tail. (Pearson Air Museum.)

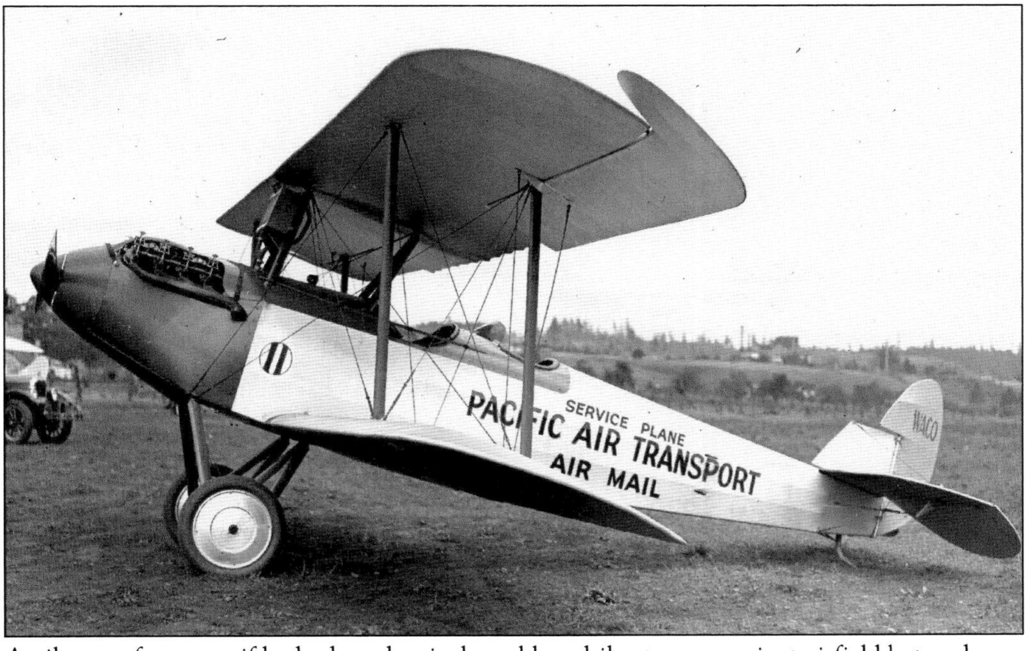

A pilot was fortunate if he had mechanical troubles while at a convenient airfield but such was rarely the case. More often than not, he would be forced to set down on an open field. Pacific Air Transport kept this Waco as a service plane, and mechanic pilots such as Frank Anderlion would fly in with parts to make repairs. (Pearson Air Museum.)

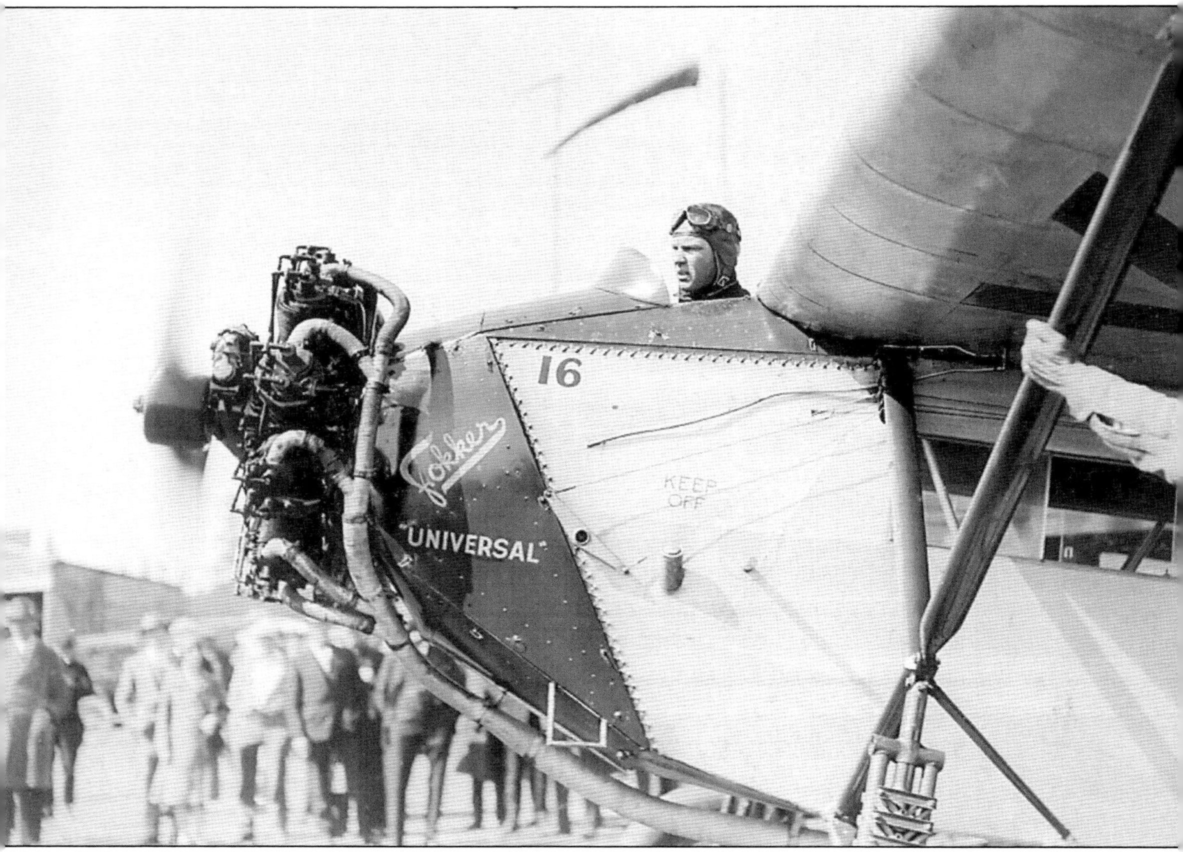

Pacific Air Transport pilot Vern Bookwalter prepares to take off in one of PAT's newest aircraft, the Fokker Universal. An increasing number of people sought passage on the airmail flights, and the Fokker had room to accommodate both cargo and four to six passengers in the cabin, while the pilot remained seated in the open cockpit behind the motor. (Pearson Air Museum.)

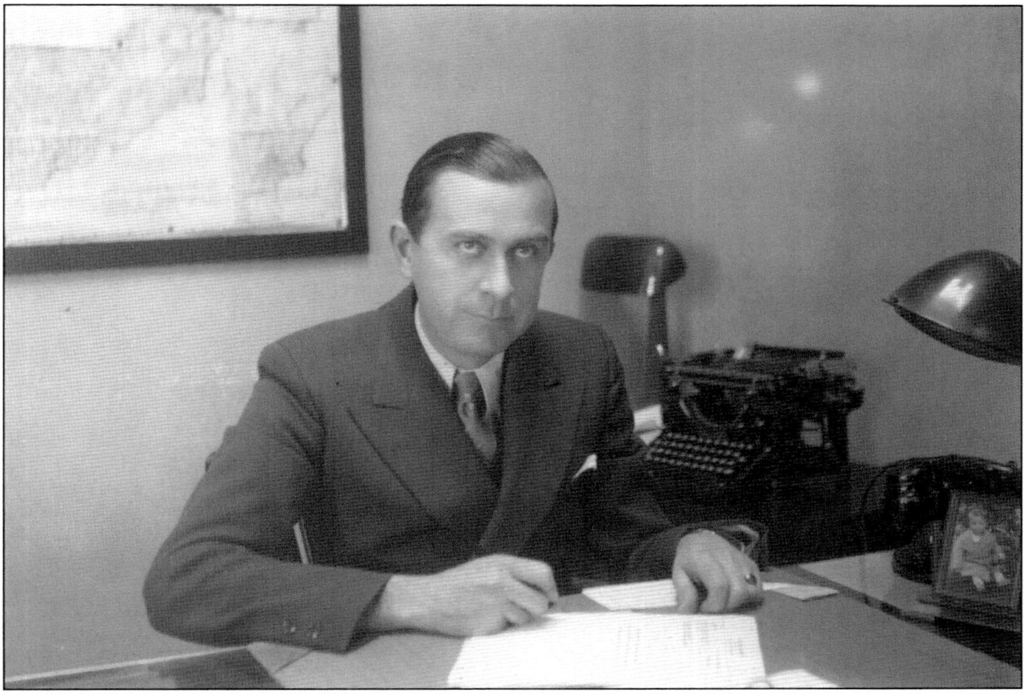

Seely Hall, known as the father of aviation in southern Oregon, had the region's first airplane, the *Mayfly*, in 1919 and was one of the earliest managers with Pacific Air Transport. He ended his career as United Airlines' vice president for ground operations. (SOHS No. 17033.)

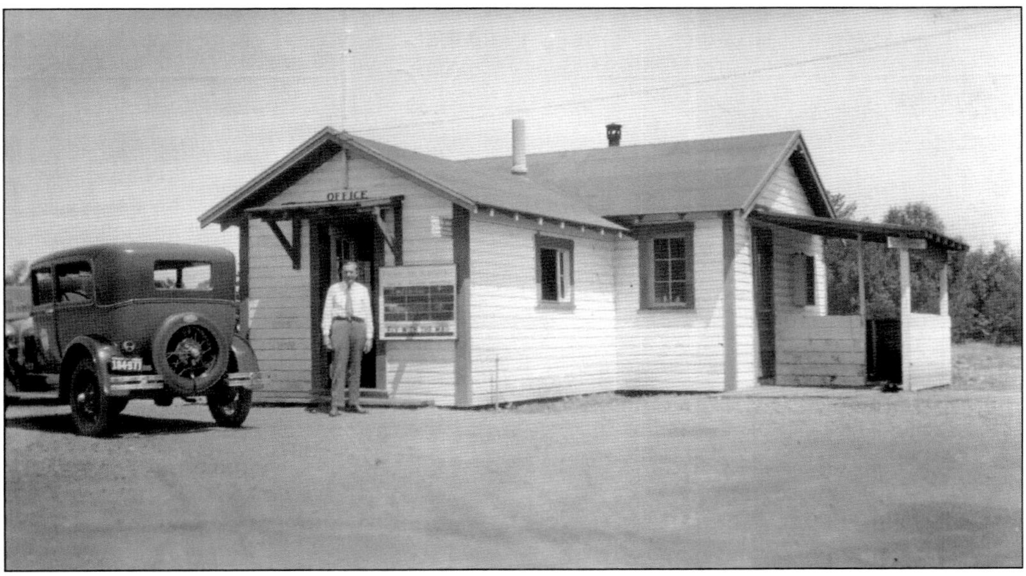

Pacific Air Transport built a small office adjacent to their hangar at Newell Barber Field. It housed the local office, space for the few passengers beginning to take to the air, and a place for overnighting pilots to catch up on their sleep. In the doorway is Seely Hall, the local field manager. (SOHS No. 957.)

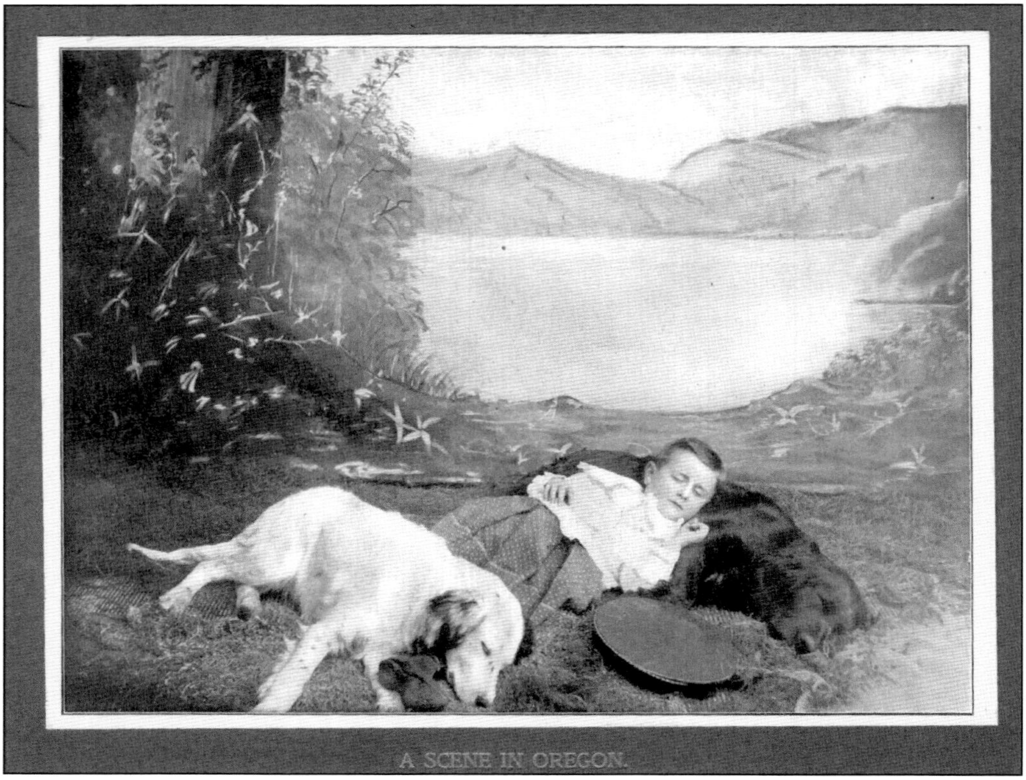

When this photograph of young Seely Hall resting his head on his dog, Jinks, was taken at the close of the 19th century, the technology that dominated his adult life was still the stuff of fiction. (SOHS No. 20732.)

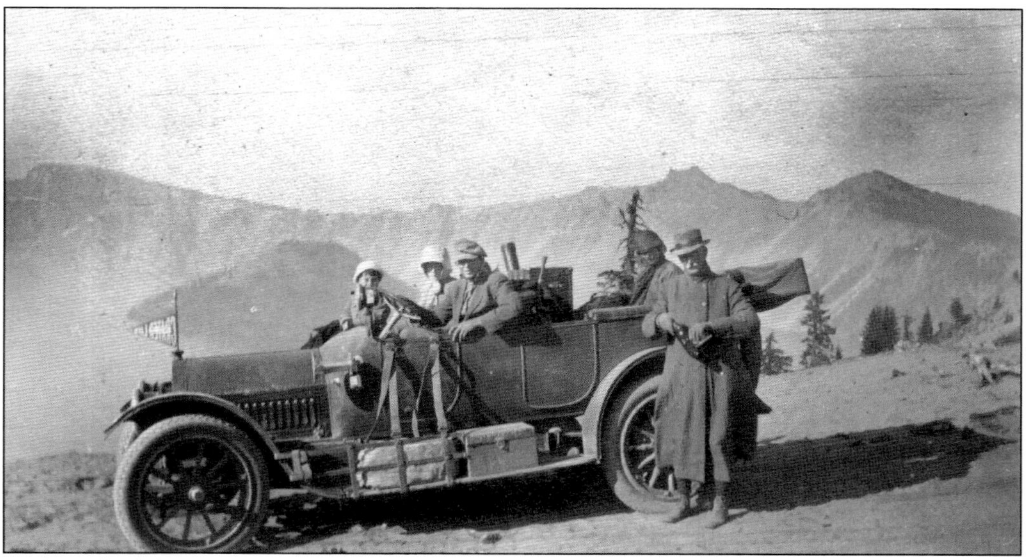

Seely Hall's father operated the Hall Taxi Company in Medford. In 1914, the Halls began offering round-trip automobile excursions to Crater Lake for $13, with Seely behind the wheel on the still primitive roads. (SOHS No.13121.)

Central Point resident Margaret Scherer booked passage to fly from Newell Barber Field to Oakland, California, on August 3, 1929. Her surviving ticket shows a total fare of $30, signed by Pacific Air Transport's Medford manager, Seely Hall. (SOHS Manuscript.)

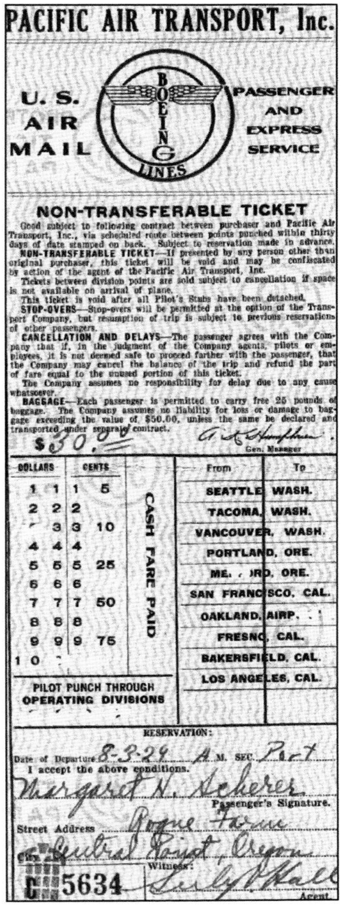

PACIFIC AIR TRANSPORT, Inc.

U. S. AIR MAIL — BOEING LINES — PASSENGER AND EXPRESS SERVICE

NON-TRANSFERABLE TICKET

Good subject to following contract between purchaser and Pacific Air Transport, Inc., via scheduled route between points punched within thirty days of date stamped on back. Subject to reservation made in advance.

NON-TRANSFERABLE TICKET—If presented by any person other than original purchaser, this ticket will be void and may be confiscated by action of the agent of the Pacific Air Transport, Inc.

Tickets between division points are sold subject to cancellation if space is not available on arrival of plane.

This ticket is void after all Pilot's Stubs have been detached.

STOP-OVERS—Stop-overs will be permitted at the option of the Transport Company, but resumption of trip is subject to previous reservations of other passengers.

CANCELLATION AND DELAYS—The passenger agrees with the Company that if, in the judgment of the Company agents, pilots or employees, it is not deemed safe to proceed farther with the passenger, that the Company may cancel the balance of the trip and refund the part of fare equal to the unused portion of this ticket.

The Company assumes no responsibility for delay due to any cause whatsoever.

BAGGAGE—Each passenger is permitted to carry free 25 pounds of baggage. The Company assumes no liability for loss or damage to baggage exceeding the value of $50.00, unless the same be declared and transported under separate contract.

$30.00

Gen. Manager

DOLLARS		CENTS		CASH FARE PAID	From	To
1	1	1	5		SEATTLE	WASH.
2	2	2			TACOMA,	WASH.
	3	3	10		VANCOUVER,	WASH.
4	4	4			PORTLAND,	ORE.
5	5	5	25		MEDFORD,	ORE.
6	6	6			SAN FRANCISCO,	CAL.
7	7	7	50		OAKLAND,	AIRP.
8	8	8			FRESNO,	CAL.
9	9	9	75		BAKERSFIELD,	CAL.
10					LOS ANGELES,	CAL.

PILOT PUNCH THROUGH OPERATING DIVISIONS

RESERVATION:

Date of Departure 8-3-29 A.M. SEC. Port

I accept the above conditions.

Margaret N. Scherer Passenger's Signature.

Street Address Rogue Farm

City Central Point, Oregon

Witness: Seely Hall Agent.

C 5634

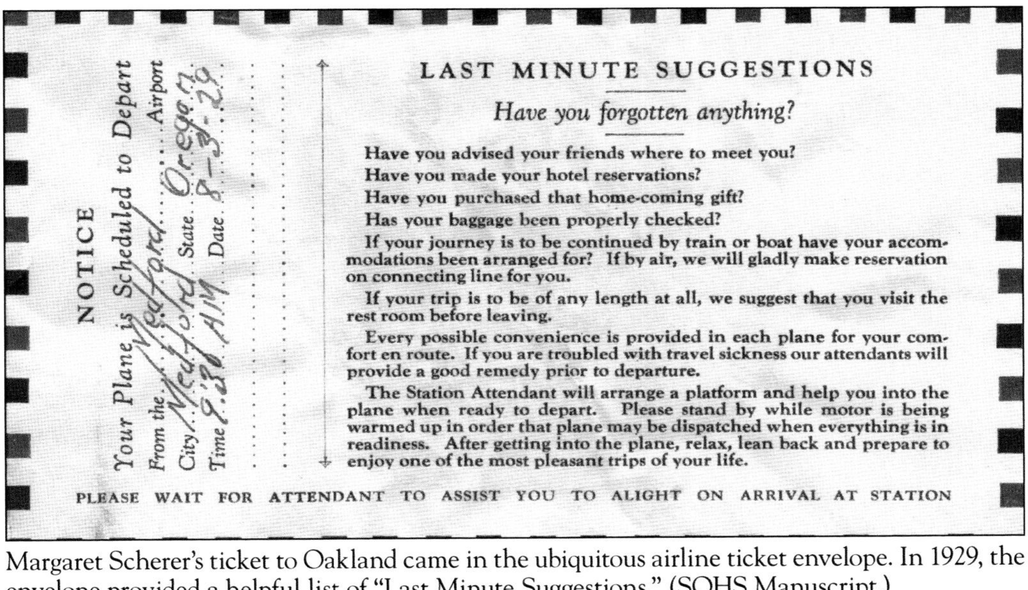

NOTICE

Your Plane is Scheduled to Depart

From the Medford Airport

City Medford State Oregon

Time 9:30 AM Date 8-3-29

LAST MINUTE SUGGESTIONS

Have you forgotten anything?

Have you advised your friends where to meet you?

Have you made your hotel reservations?

Have you purchased that home-coming gift?

Has your baggage been properly checked?

If your journey is to be continued by train or boat have your accommodations been arranged for? If by air, we will gladly make reservation on connecting line for you.

If your trip is to be of any length at all, we suggest that you visit the rest room before leaving.

Every possible convenience is provided in each plane for your comfort en route. If you are troubled with travel sickness our attendants will provide a good remedy prior to departure.

The Station Attendant will arrange a platform and help you into the plane when ready to depart. Please stand by while motor is being warmed up in order that plane may be dispatched when everything is in readiness. After getting into the plane, relax, lean back and prepare to enjoy one of the most pleasant trips of your life.

PLEASE WAIT FOR ATTENDANT TO ASSIST YOU TO ALIGHT ON ARRIVAL AT STATION

Margaret Scherer's ticket to Oakland came in the ubiquitous airline ticket envelope. In 1929, the envelope provided a helpful list of "Last Minute Suggestions." (SOHS Manuscript.)

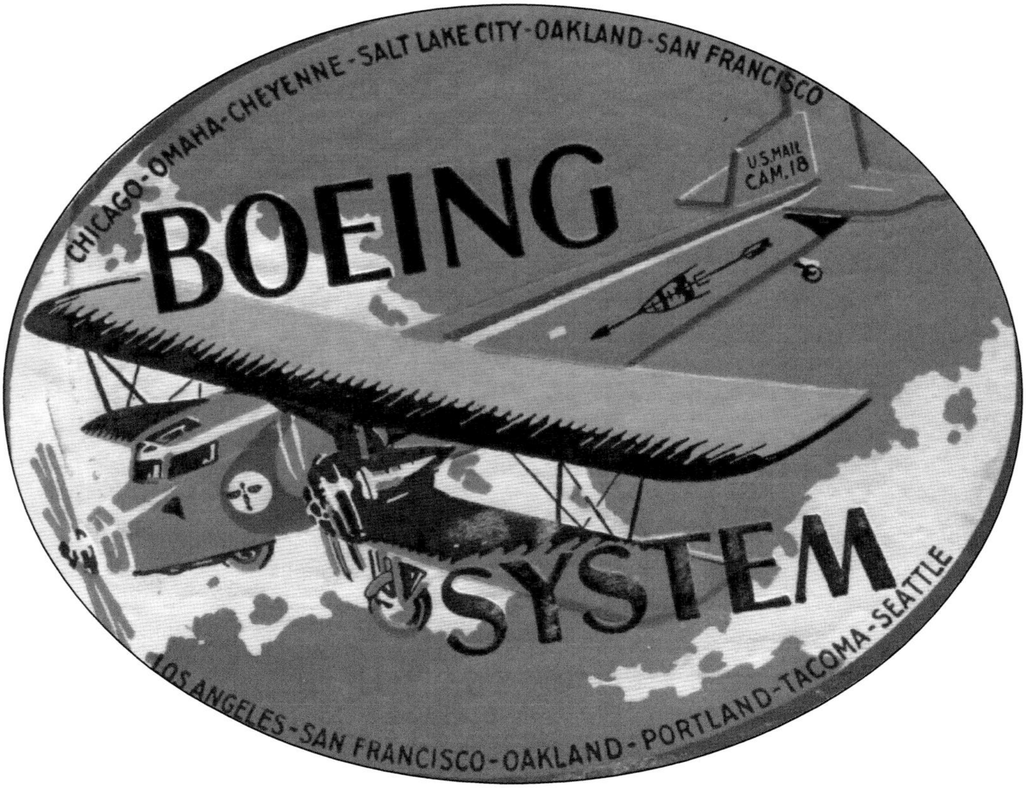

Luggage stickers are a thing of the past today, but during the 1920s and 1930s they were commonplace. This colorful Boeing System luggage sticker highlights Boeing's Contract Airmail Route (CAM) 18 between San Francisco and Chicago, as well as CAM 8, the West Coast route operated by subsidiary Pacific Air Transport. (SOHS Manuscript.)

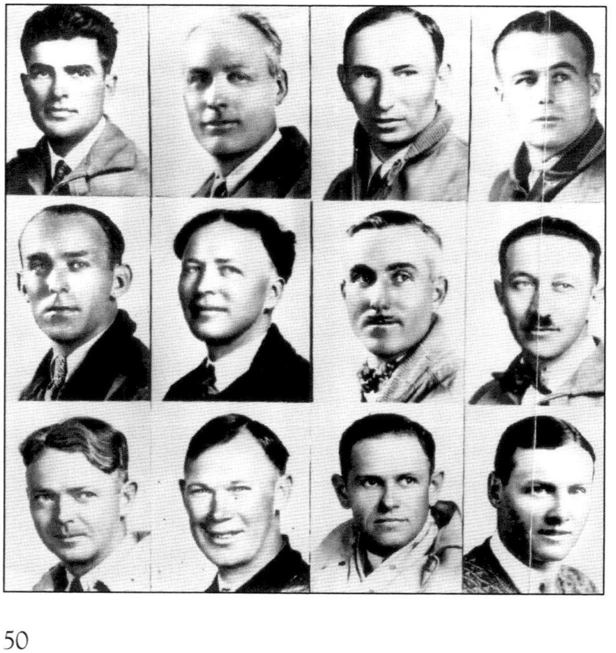

Twelve of the Pacific Air Transport pilots are, regretfully, unidentified individually in this photograph that once belonged to Seely Hall. (SOHS No. 9595.)

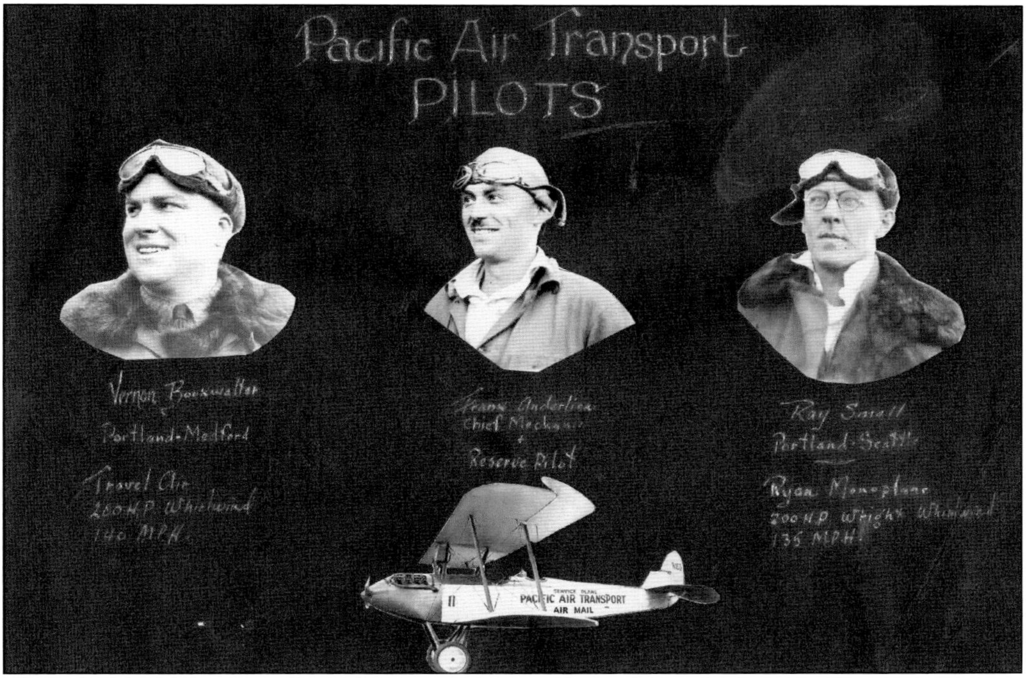

This scrapbook page features the northwest pilots of Pacific Air Transport. Vern Bookwalter, left, flew the Medford-Vancouver/Portland leg; Frank Anderlion was chief mechanic and reserve pilot; and Ray Small flew between Seattle and Vancouver/Portland. (Pearson Air Museum.)

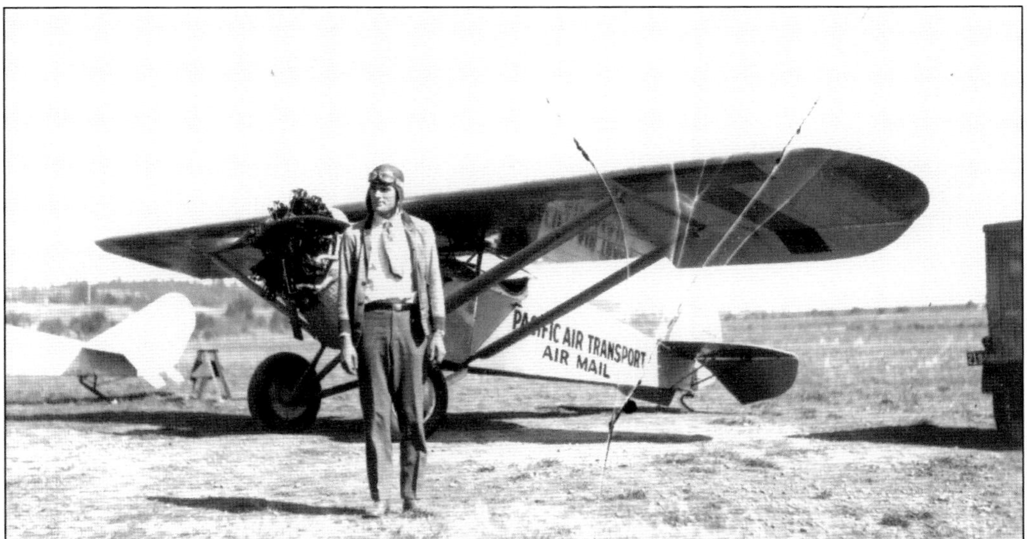

Veteran Pacific Air Transport pilot Art Starbuck was a familiar site at Newell Barber Field. He often flew the Medford-Oakland leg of CAM 8 and was the pilot who taught Harold Sander, the first local flight school proprietor, how to fly. On May 6, 1931, while piloting a night flight out of Burbank, Starbuck flew into a mountainside and was killed. It was his wish that his remains be scattered over the rugged Siskiyou Mountains along his favorite section of CAM 8, the Medford to Oakland leg. (Oregon Historical Society.)

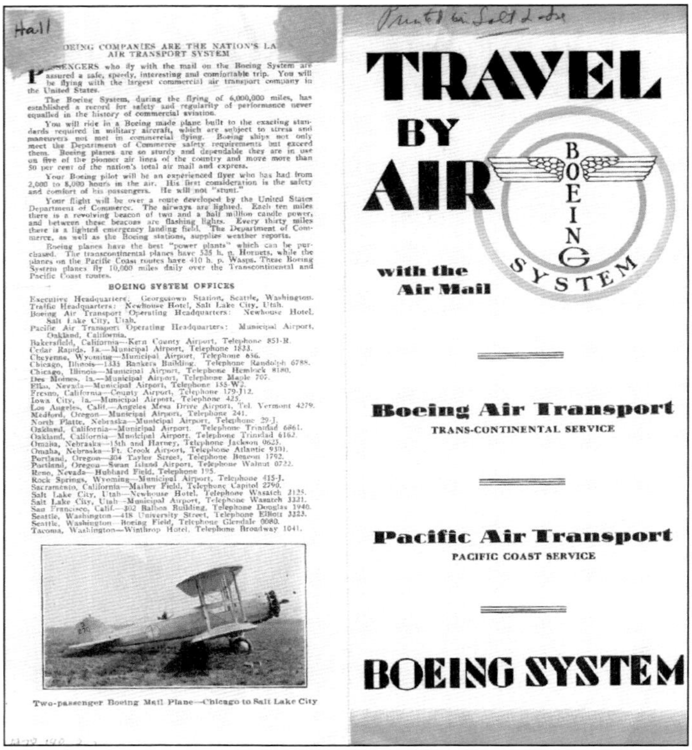

Hall

OEING COMPANIES ARE THE NATION'S LA
AIR TRANSPORT SYSTEM

PASSENGERS who fly with the mail on the Boeing System are assured a safe, speedy, interesting and comfortable trip. You will be flying with the largest commercial air transport company in the United States.

The Boeing System, during the flying of 6,000,000 miles, has established a record for safety and regularity of performance never equalled in the history of commercial aviation.

You will ride in a Boeing made plane built to the exacting standards required in military aircraft, which are subject to stress and maneuvers not met in commercial flying. Boeing ships not only meet the Department of Commerce safety requirements but exceed them. Boeing planes are so sturdy and dependable they are in use on five of the pioneer air lines of the country and move more than 50 per cent of the nation's total air mail and express.

Your Boeing pilot will be an experienced flyer who has had from 2,000 to 8,000 hours in the air. His first consideration is the safety and comfort of his passengers. He will not "stunt."

Your flight will be over a route developed by the United States Department of Commerce. The airways are lighted. Each ten miles there is a revolving beacon of two and a half million candle power, and between these beacons are flashing lights. Every thirty miles there is a lighted emergency landing field. The Department of Commerce, as well as the Boeing stations, supplies weather reports.

Boeing planes have the best "power plants" which can be purchased. The transcontinental planes have 525 h. p. Hornets, while the planes on the Pacific Coast routes have 410 h. p. Wasps. These Boeing System planes fly 10,000 miles daily over the Transcontinental and Pacific Coast routes.

BOEING SYSTEM OFFICES

Executive Headquarters. Georgetown Station, Seattle, Washington.
Traffic Headquarters: Newhouse Hotel, Salt Lake City, Utah.
Boeing Air Transport Operating Headquarters: Newhouse Hotel, Salt Lake City, Utah.
Pacific Air Transport Operating Headquarters: Municipal Airport, Oakland, California.
Bakersfield, California—Kern County Airport, Telephone 851-R.
Cedar Rapids, Ia.—Municipal Airport, Telephone 1833.
Cheyenne, Wyoming—Municipal Airport, Telephone 656.
Chicago, Illinois—1335 Bankers Building. Telephone Randolph 6788.
Chicago, Illinois—Municipal Airport, Telephone Hemlock 8180.
Des Moines, Ia.—Municipal Airport, Telephone Maple 707.
Elko, Nevada—Municipal Airport, Telephone 155-W2.
Fresno, California—County Airport, Telephone 179-J12.
Iowa City, Ia.—Municipal Airport, Telephone 425.
Los Angeles, Calif.—Angeles Mesa Drive Airport, Tel. Vermont 4279.
Medford, Oregon—Municipal Airport, Telephone 241.
North Platte, Nebraska—Municipal Airport, Telephone 29-J.
Oakland, California—Municipal Airport. Telephone Trinidad 6861.
Oakland, California—Municipal Airport, Telephone Trinidad 6162.
Omaha, Nebraska—15th and Harney, Telephone Jackson 0625.
Omaha, Nebraska—Ft. Crook Airport, Telephone Atlantic 9301.
Portland, Oregon—304 Taylor Street, Telephone Beacon 1792.
Portland, Oregon—Swan Island Airport, Telephone Walnut 0722.
Reno, Nevada—Hubbard Field, Telephone 195.
Rock Springs, Wyoming—Municipal Airport, Telephone 415-J.
Sacramento, California—Mather Field, Telephone Capitol 2790.
Salt Lake City, Utah—Newhouse Hotel, Telephone Wasatch 2125.
Salt Lake City, Utah—Municipal Airport, Telephone Wasatch 3321.
San Francisco, Calif.—302 Balboa Building, Telephone Douglas 1940.
Seattle, Washington—418 University Street, Telephone Elliott 3323.
Seattle, Washington—Boeing Field, Telephone Glendale 0080.
Tacoma, Washington—Winthrop Hotel, Telephone Broadway 1041.

Two-passenger Boeing Mail Plane—Chicago to Salt Lake City

TRAVEL BY AIR

BOEING SYSTEM

with the Air Mail

Boeing Air Transport
TRANS-CONTINENTAL SERVICE

Pacific Air Transport
PACIFIC COAST SERVICE

BOEING SYSTEM

Improvements in the size and performance of the newer generation of aircraft threatened Newell Barber Field with obsolescence, and some of the insurance underwriters were beginning to condemn the old field. Pacific Air Transport, now operated by Boeing Air Transport, worked with the City to solve the dilemma. (SOHS Manuscript.)

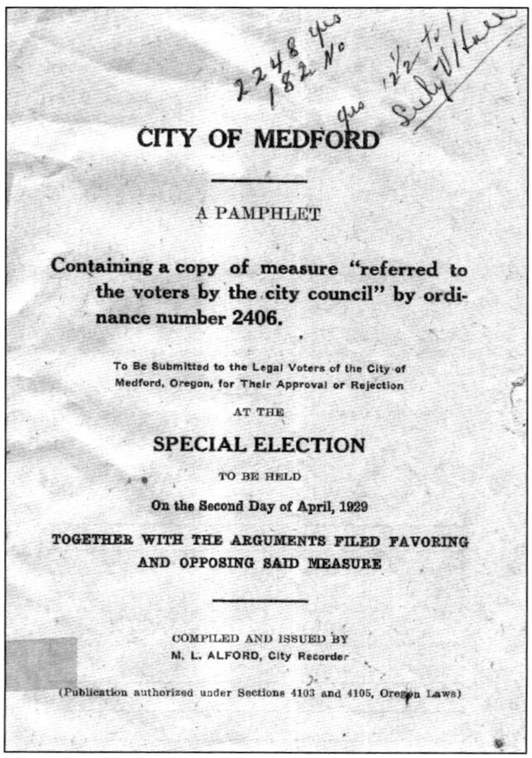

CITY OF MEDFORD

A PAMPHLET

Containing a copy of measure "referred to the voters by the city council" by ordinance number 2406.

To Be Submitted to the Legal Voters of the City of Medford, Oregon, for Their Approval or Rejection

AT THE

SPECIAL ELECTION

TO BE HELD

On the Second Day of April, 1929

TOGETHER WITH THE ARGUMENTS FILED FAVORING AND OPPOSING SAID MEASURE

COMPILED AND ISSUED BY
M. L. ALFORD, City Recorder

(Publication authorized under Sections 4103 and 4105, Oregon Laws)

Aviation enthusiasts wanted to keep Medford "an Air-Minded City" and convinced the City to submit a bond proposal for a new airport to the voters. This copy of the proposal is annotated with the results by Pacific Air Transport manager Seely Hall: 2,248 votes in favor, 182 against, a margin of 12.5 to 1. Work on the new airfield began almost at once. (SOHS Manuscript.)

Four

The New Municipal Airport 1929–1945

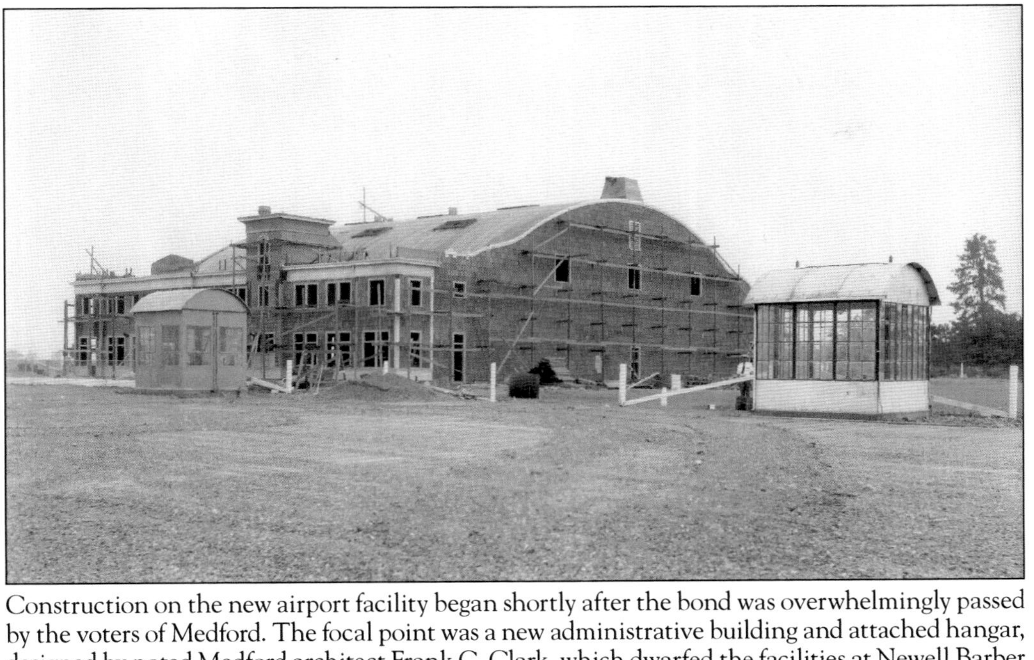

Construction on the new airport facility began shortly after the bond was overwhelmingly passed by the voters of Medford. The focal point was a new administrative building and attached hangar, designed by noted Medford architect Frank C. Clark, which dwarfed the facilities at Newell Barber Field. (Rogue Valley International-Medford Airport.)

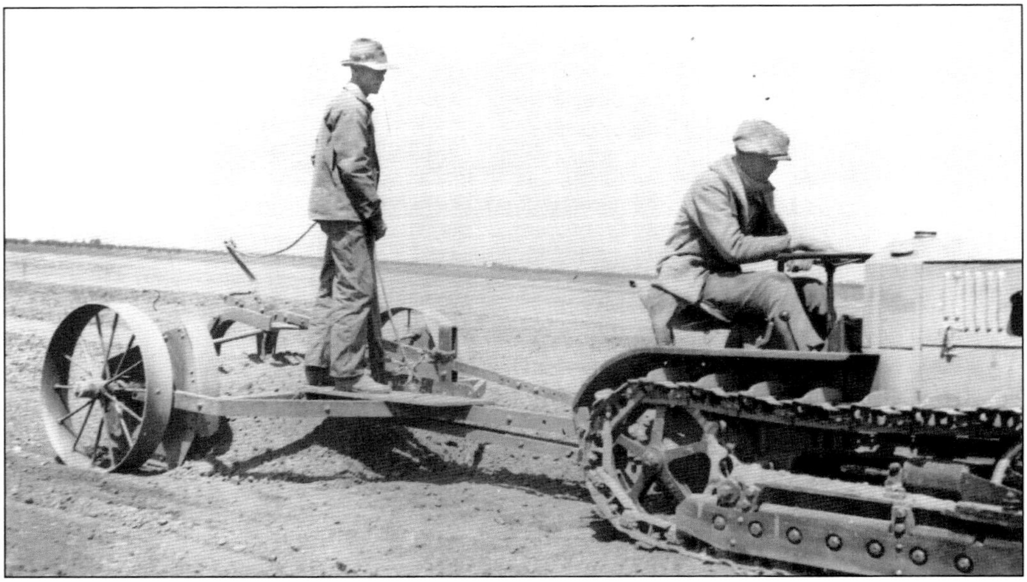

Little time was wasted after the passage of the new airport bond measure. Construction crews soon began grading for the 3,600-foot gravel runway north of town. (SOHS No. 20707.)

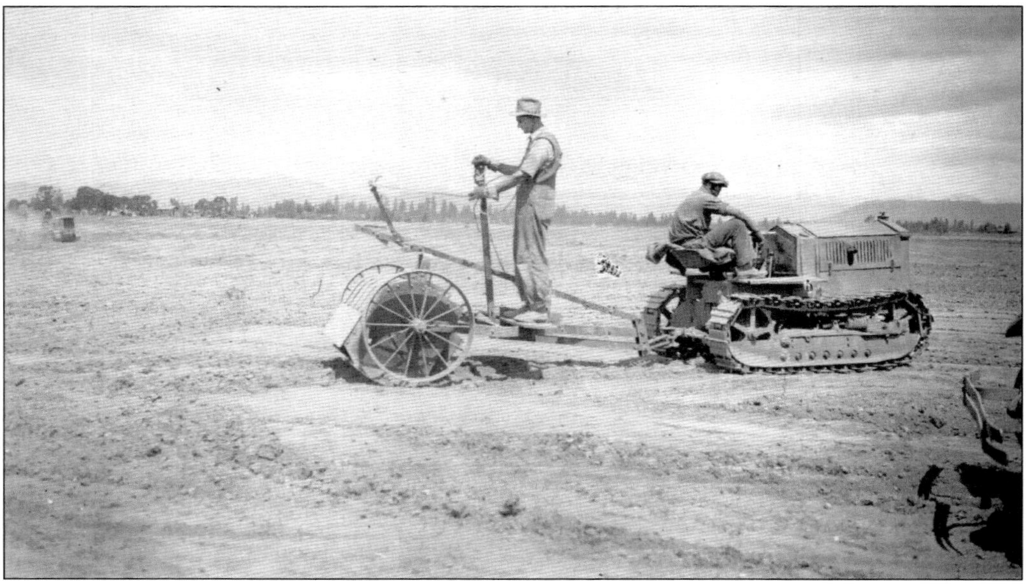

Extensive grading was required not only to ensure a level landing strip, but to allow for proper drainage of the runway during the rainy northwest winters. (SOHS No. 20706.)

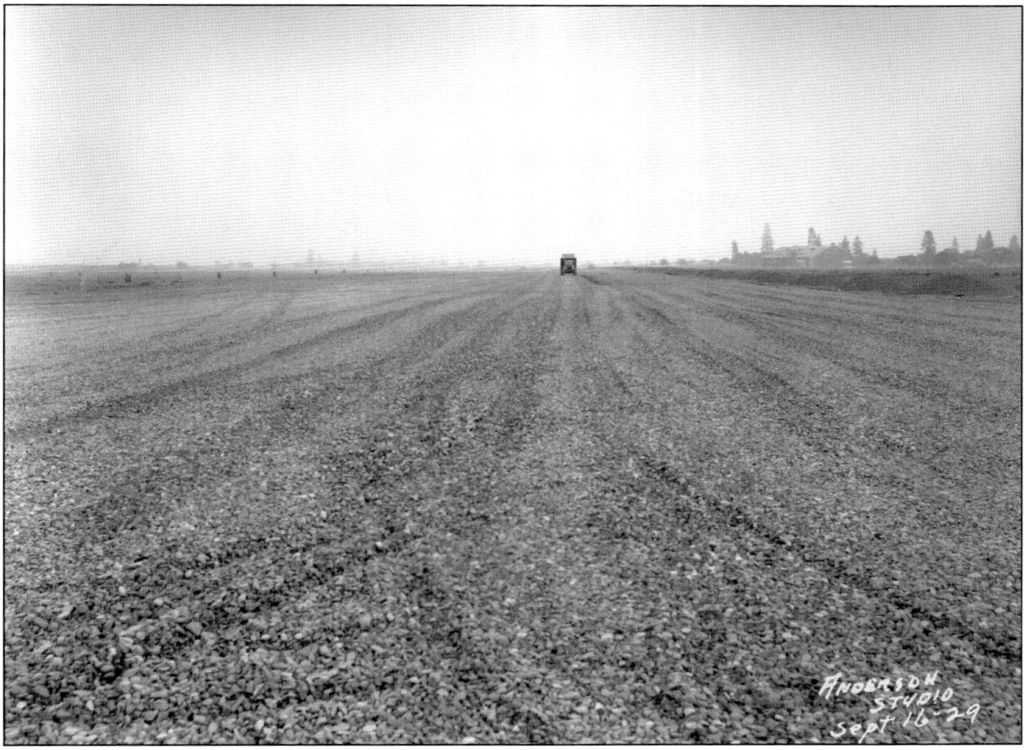

A mere five months after passage of the airport bond, the terminal nears completion and the expansive new runway is already graded and graveled. This image of a lone construction truck on the runway was taken on September 16, 1929. (Rogue Valley International-Medford Airport.)

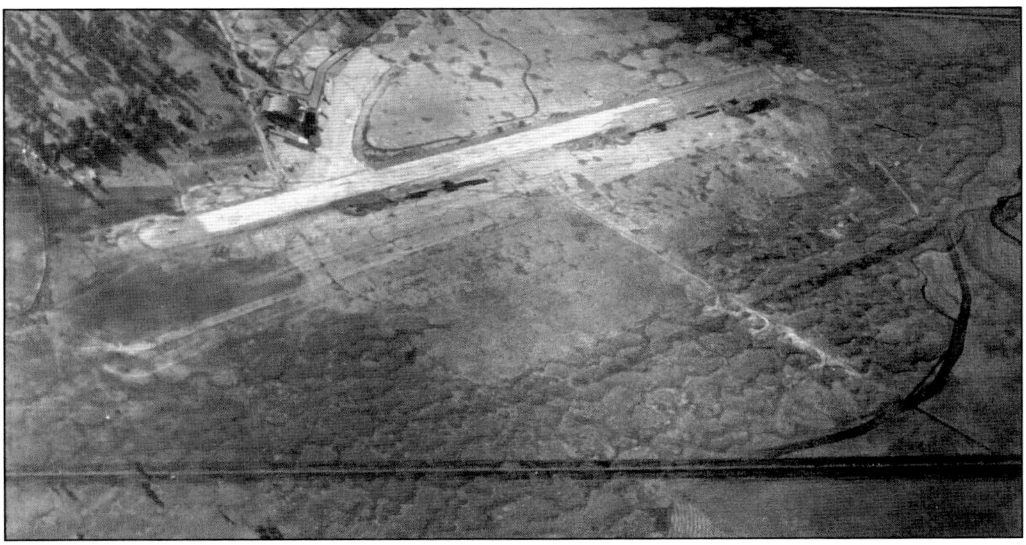

This aerial photograph of the new airport site off Biddle Road clearly shows the runway and terminal facilities. It also indicates just how much room for future expansion there was surrounding the field. The right-of-way of the Medford Logging Railroad can be seen across the bottom of the photograph. (SOHS No. 19205.)

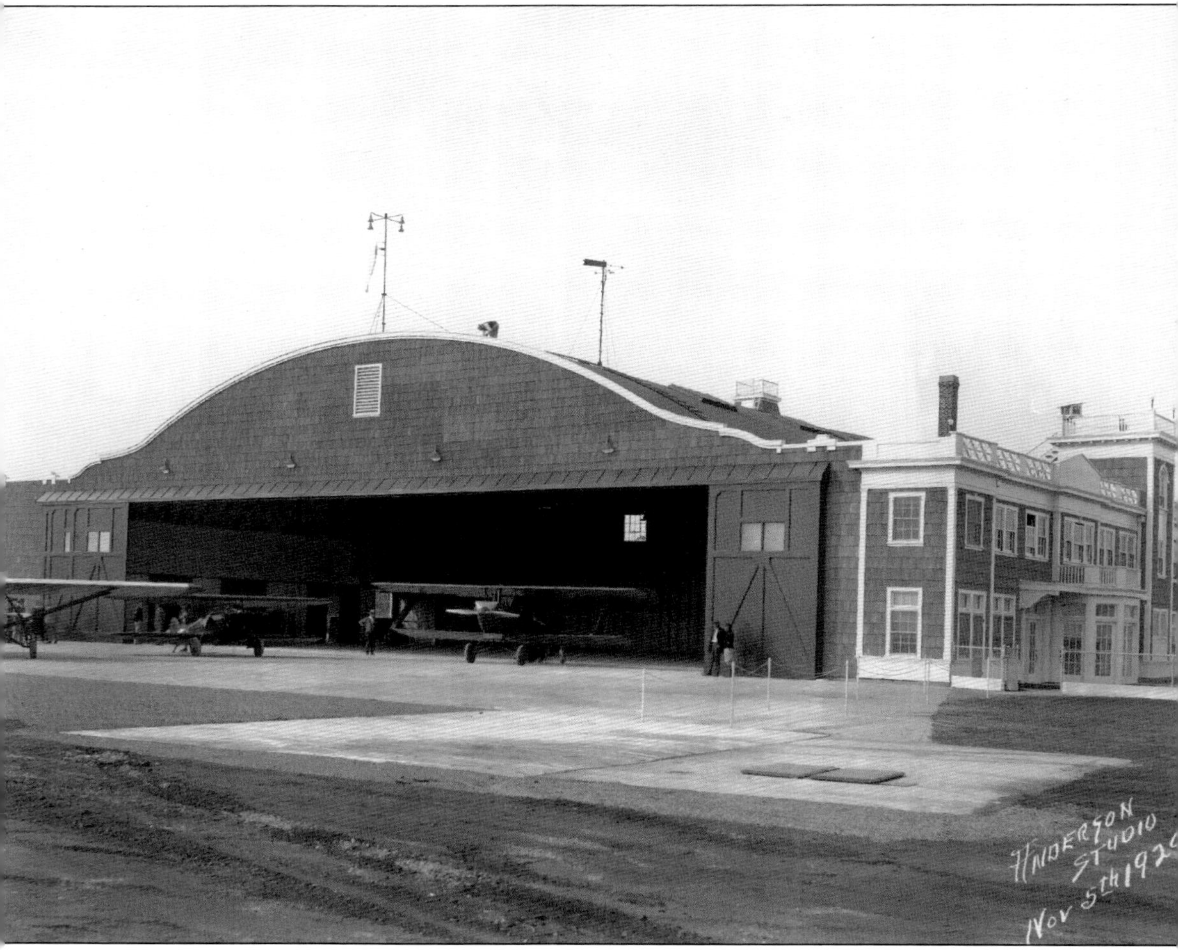

Attached to the administrative building at the new Medford Municipal Airport was a large hangar, capable of accommodating a number of aircraft. Skylights in the roof allowed natural light illumination. At the official dedication of the field in 1930, a huge dance has held in the hangar. (SOHS No. 967.)

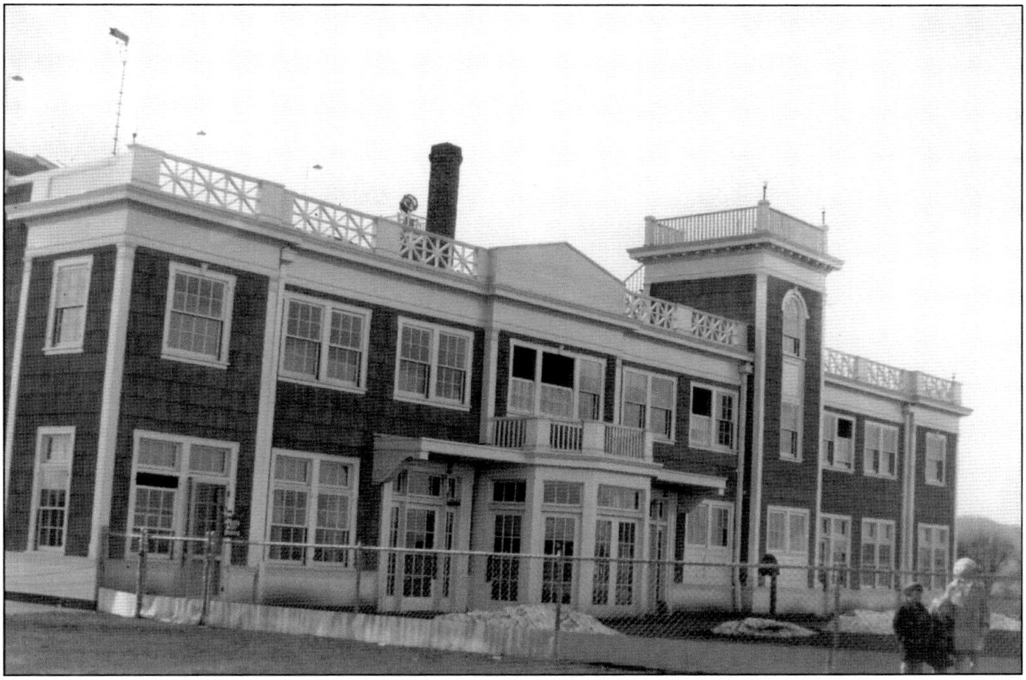

A close-up view of the newly completed Medford Municipal Airport administration and terminal facility shows some of the architectural details of Frank Clark's design. The 2.5-million candlepower beacon can be seen on the roof, and a built-in open tower provides a bird's eye view of the field. (SOHS No. 20276.)

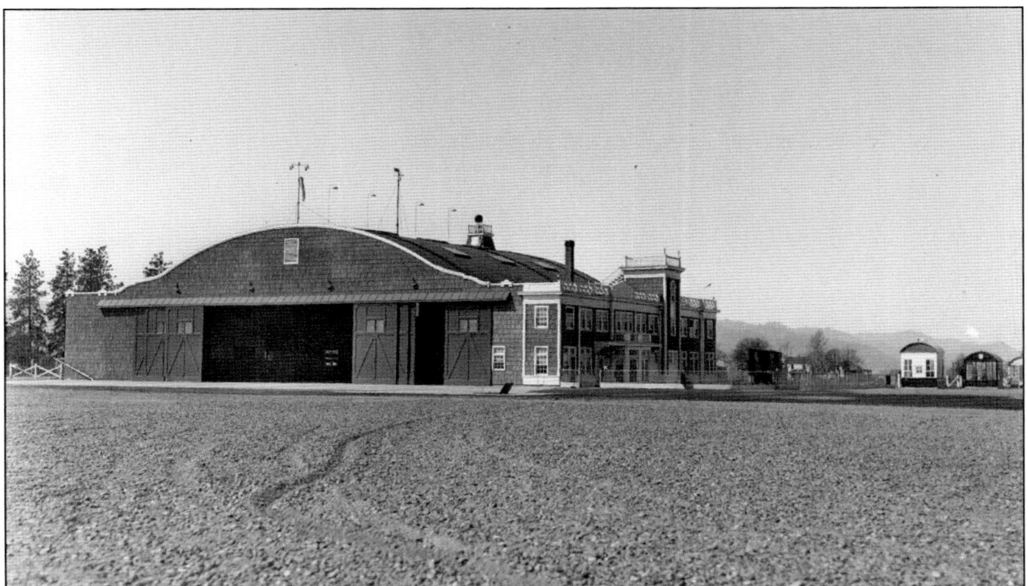

The Medford Municipal Airport administrative building and terminal was completed under budget and well before the summer of 1930's grand opening. Outside the hangar is a concrete pad and a well-developed gravel taxiway. Also in view are the three gasoline pumps for refueling aircraft. (SOHS No. 19187.)

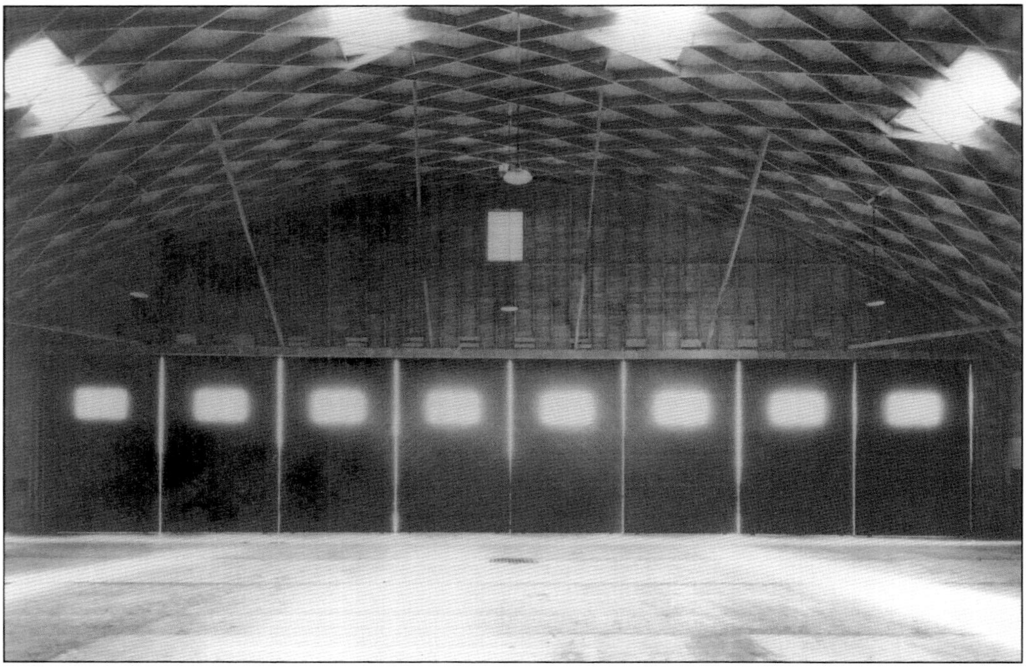
This is the interior of a large, expansive new hangar. Designed by Frank Clark, the arched wooden roof needed no interior support, providing ample room for aircraft of all sizes. (SOHS No. 967.)

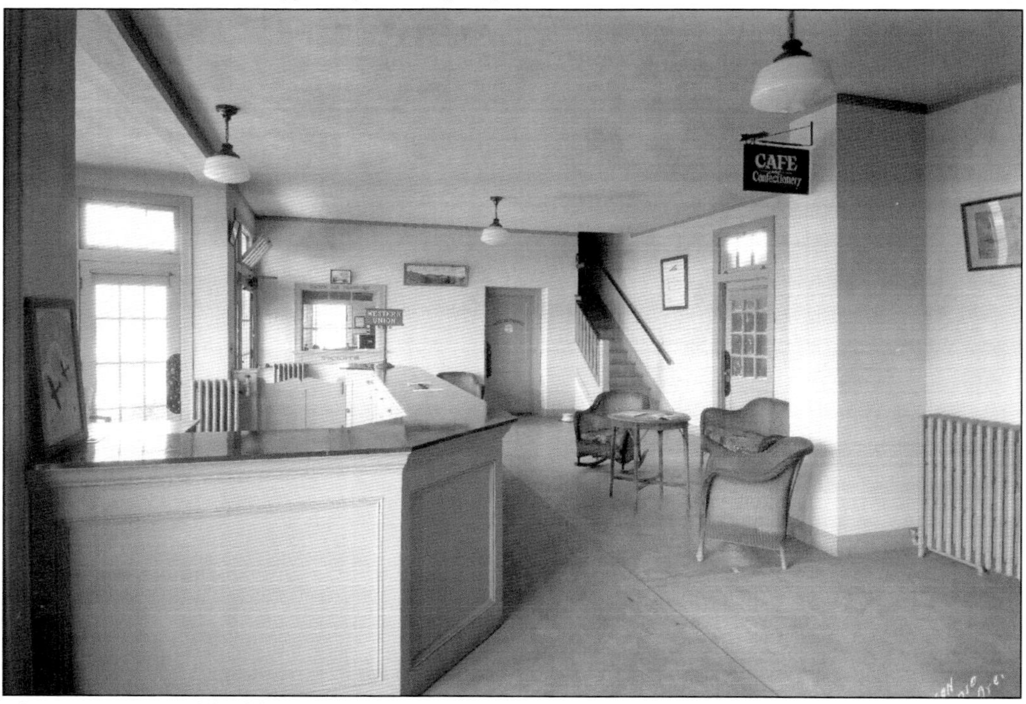

The new terminal building at the Medford Municipal Airport was a modern, state-of-the-art facility. In addition to the restaurant and offices for Pacific Air Transport, there was a comfortable lobby where passengers could relax while waiting for their flight. (SOHS No. 9605.)

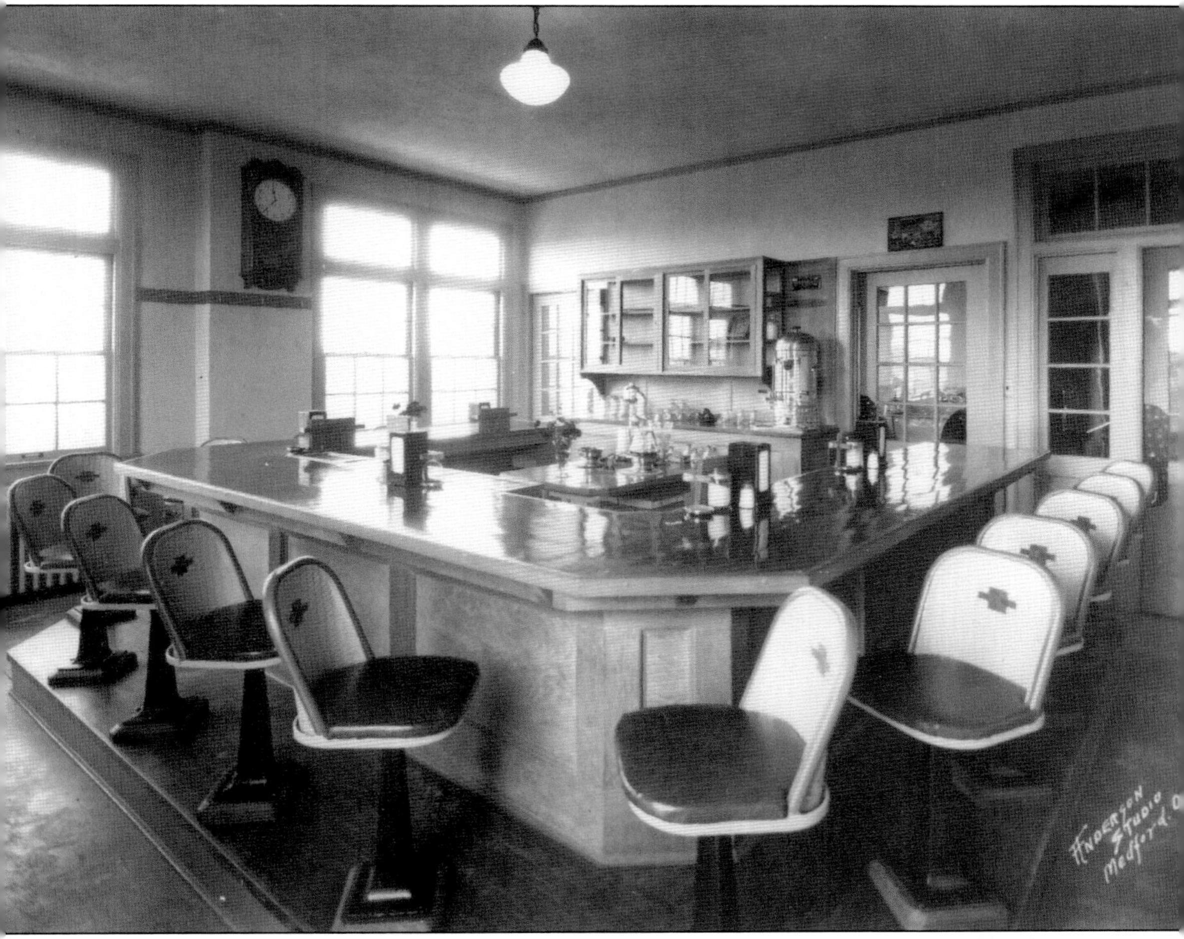

Among the many amenities of the new Medford Municipal Airport building was the coffee shop and confectionary operated by Wallace Singleton and Seath Blake, owners of the Rex Café in downtown Medford. Here passengers could grab a bite to eat between flights, staff could enjoy lunch, and residents could enjoy a meal with the ambience of aviation. (SOHS No. 3722.)

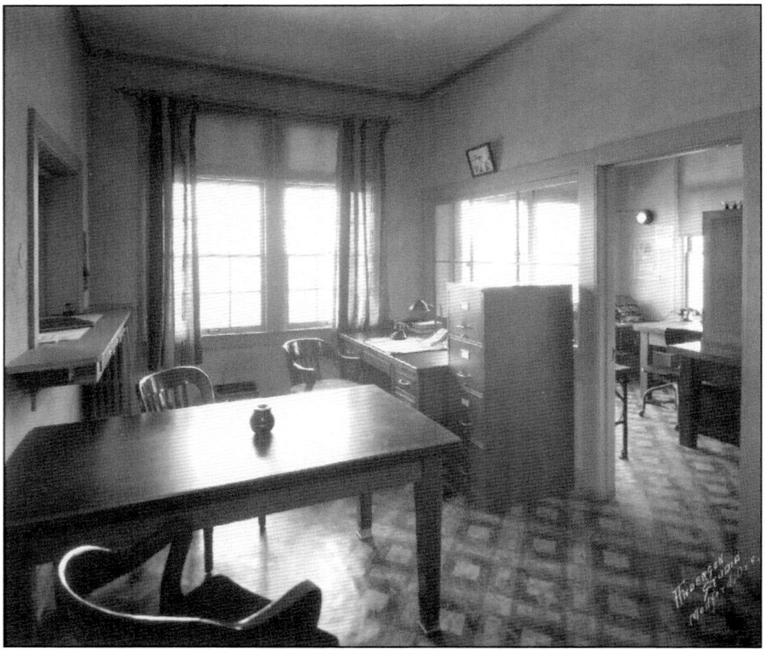

To accommodate the airmail and passenger service at the new terminal building, spacious new offices for Pacific Air Transport were included on the ground floor. The outer office had a ticket counter that opened into the main lobby. (SOHS No. 9606.)

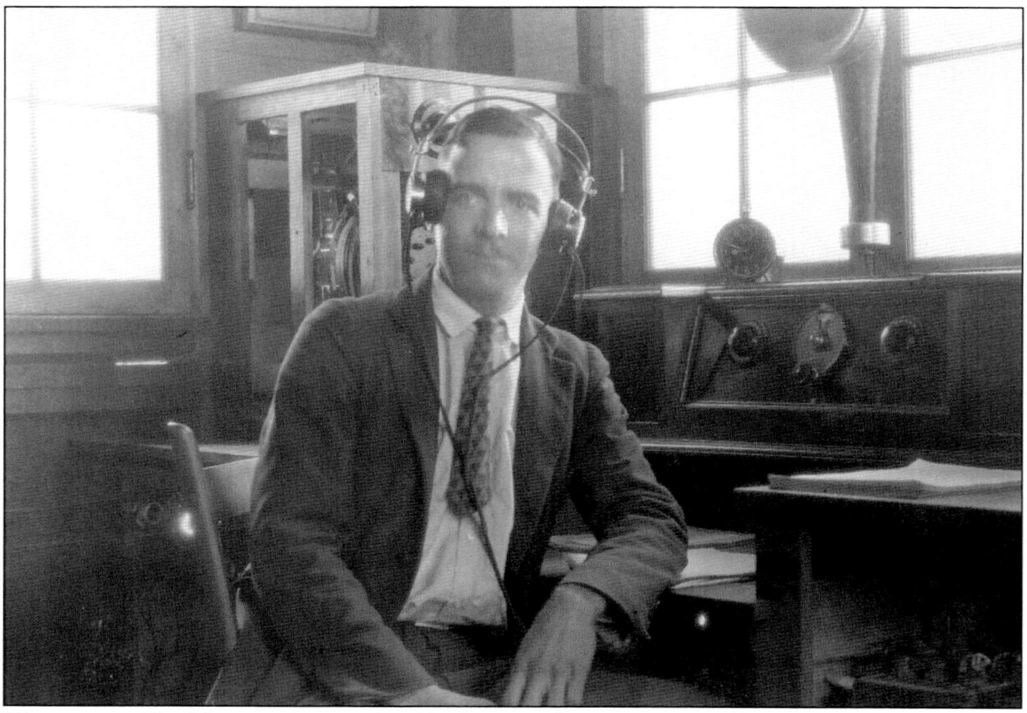

The terminal building at the new Medford airport location on Biddle Road was designed with all the latest innovations in mind. Pacific Air Transport, now operated by Boeing Air Transport, was provided a special radio room with all the latest in communications equipment. From here, PAT radio operator George Johnson could readily communicate with mail planes in flight and other ground stations. (SOHS No. 15622.)

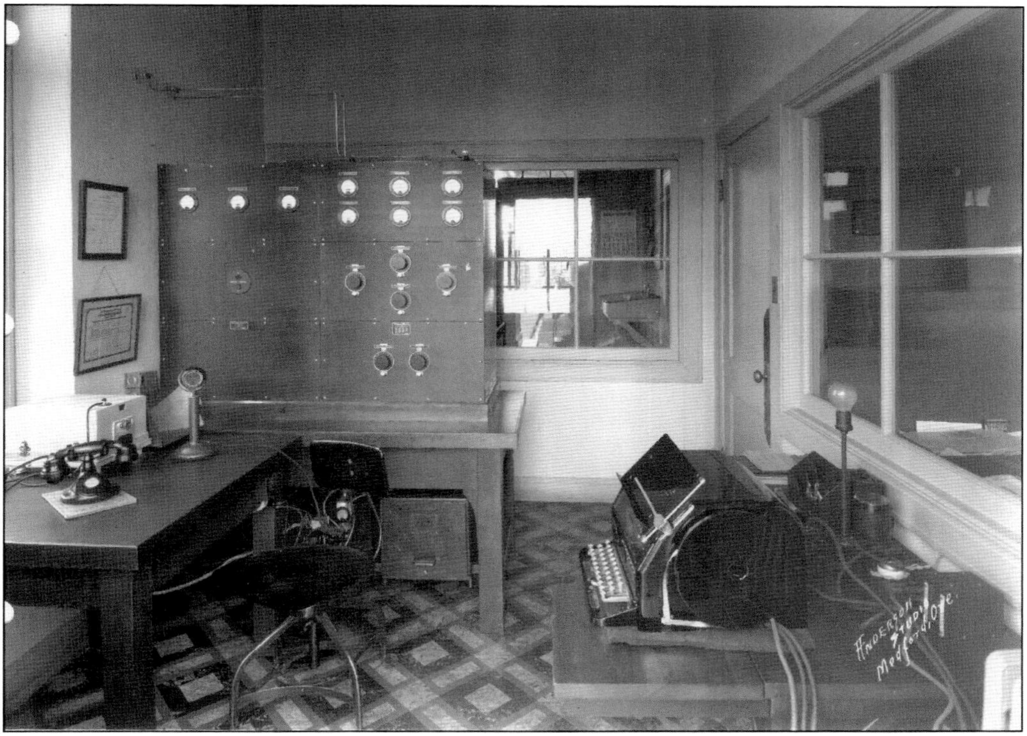

Pacific Air Transport had ample space in the new Medford terminal building to house a modern radio communications center. Ground stations could now easily maintain contact with pilots in the air, and the teletype and telephone systems allowed for instant contact with other ground stations. (SOHS No. 9599.)

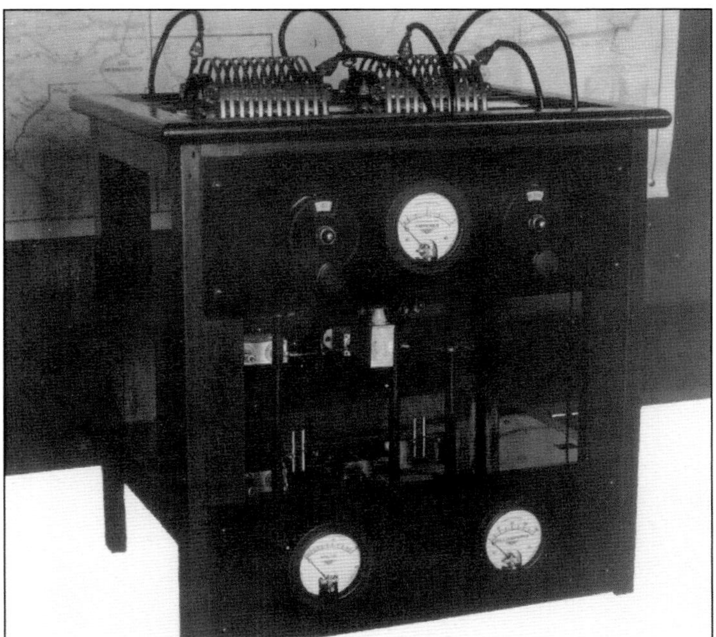

Medford's new municipal airport was equipped with all of the latest in aviation technology. Pacific Air Transport had this state-of-the-art radio-telegraph transmitter installed in its offices in 1929. (SOHS No. 9598.)

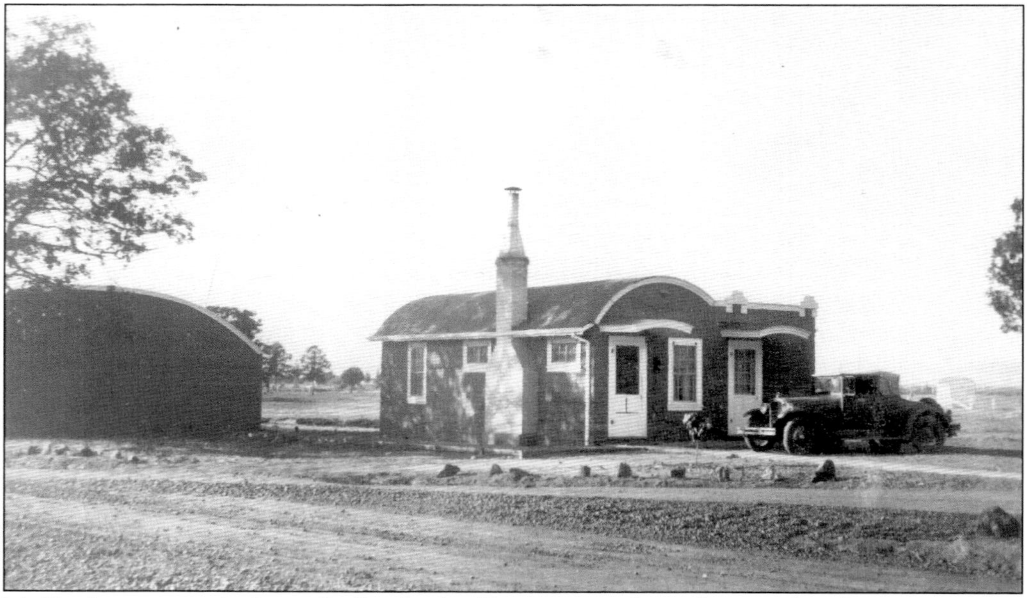

Medford was a terminal stop for Pacific Air Transport, a place where mail and passengers were transferred to a new plane with a new pilot. A comfortable pilots' quarters was built at the new airport for the comfort of crews between flights. (SOHS No. 3724.)

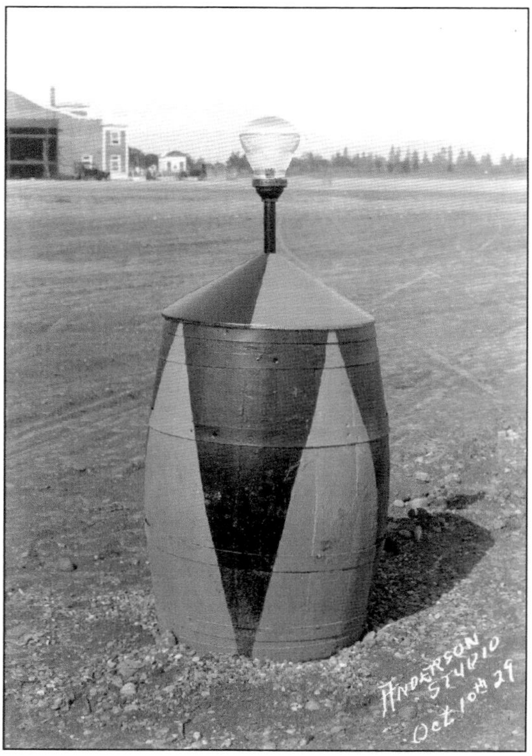

Newell Barber Field was unsuited for night flying, but the new airport had a 2.5-million candlepower beacon and runway perimeter light fixtures, designed by Seely Hall. The terminal building in the background still has scaffolding for the construction crews. (SOHS No. 20703.)

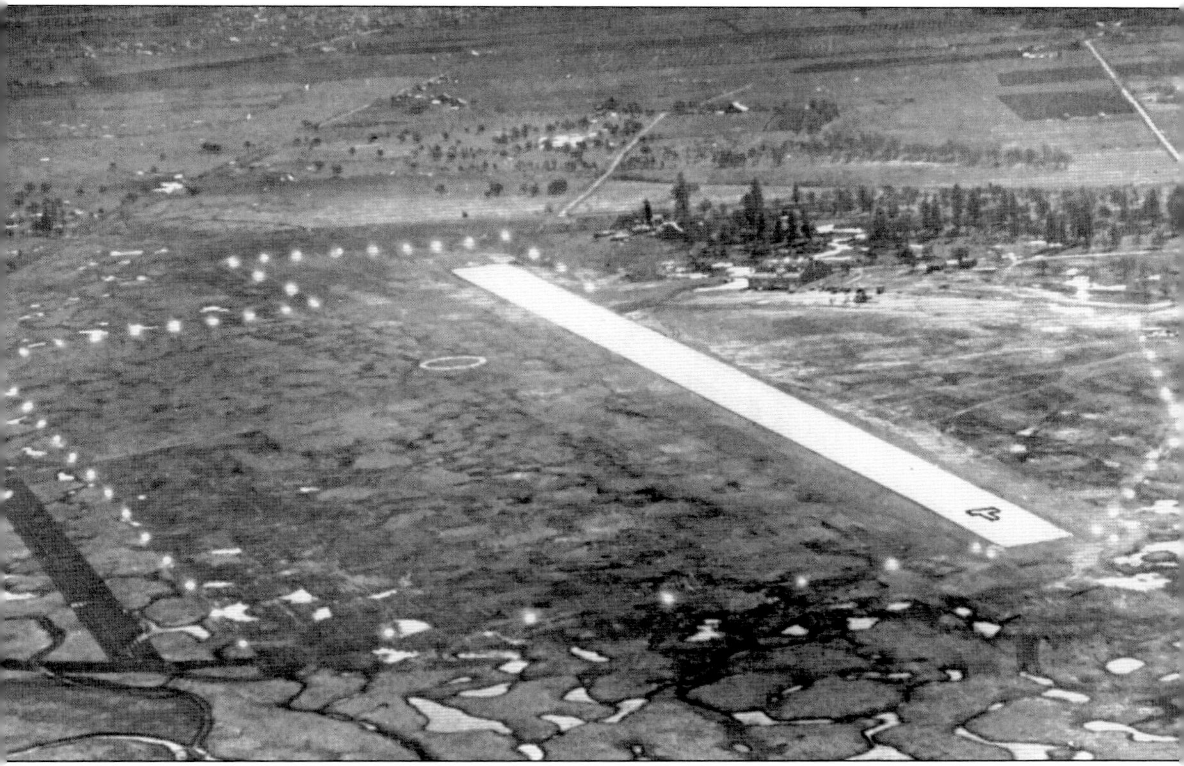

Here is an aerial view of the new Medford airport, looking to the south. This image shows the placement of the airport's perimeter lighting. (SOHS No. 8728.)

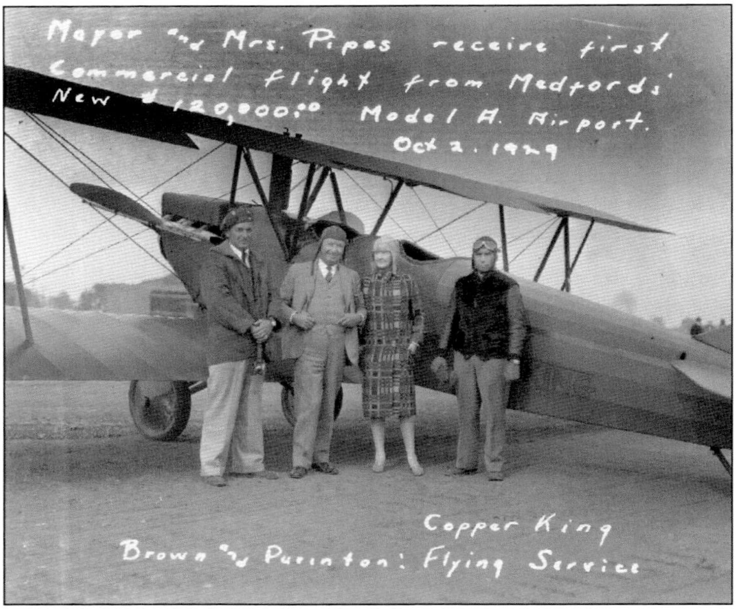

The runways at the new Medford Municipal Airport were ready for use in the fall of 1929. Shortly after the first airmail flight on October 2, Medford mayor Alfred Pipes and his wife, Maude, made a ceremonial first commercial flight with the Copper King Flying Service. It was Maude Pipes's first airplane ride. After the flight, Copper King manager W. J. Brown, the Pipes, and pilot Jimmy Purinton posed for a photograph. (SOHS No. 16669.)

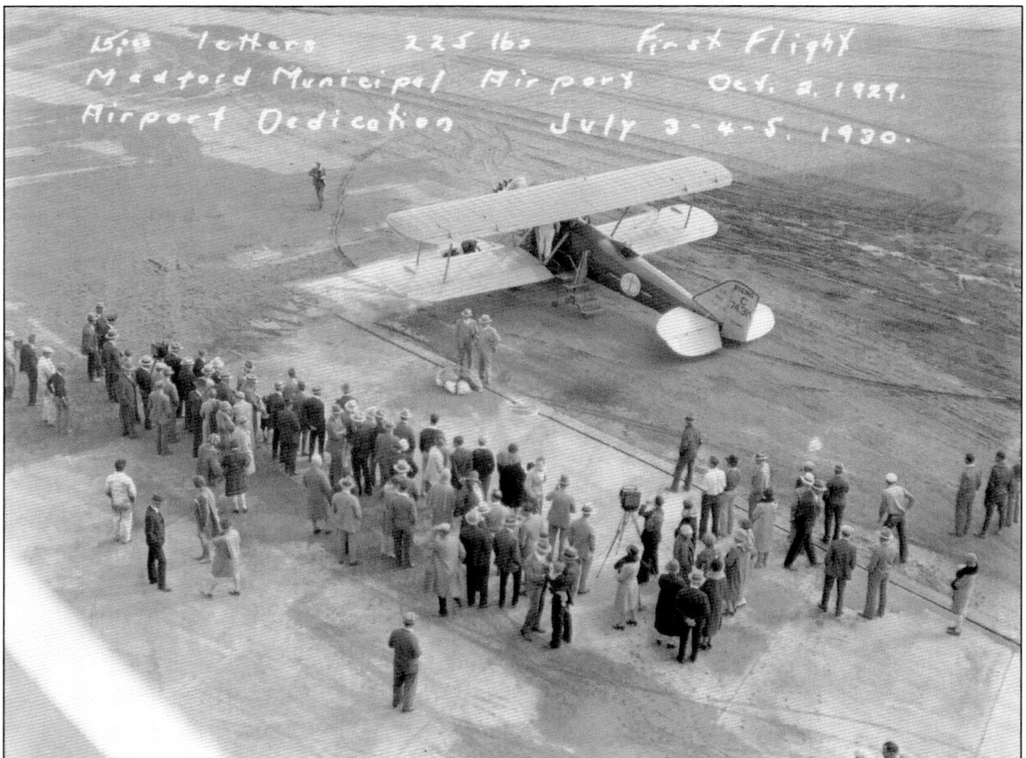

The new Medford Municipal Airport was not officially dedicated until the summer of 1930, but Pacific Air Transport began using the field in the fall of 1929. On October 2, 1929, a sizeable crowd gathered to witness the first airmail flight from the new airport. Fifteen thousand letters were loaded onto a Boeing 40-B for transport. (SOHS 16670.)

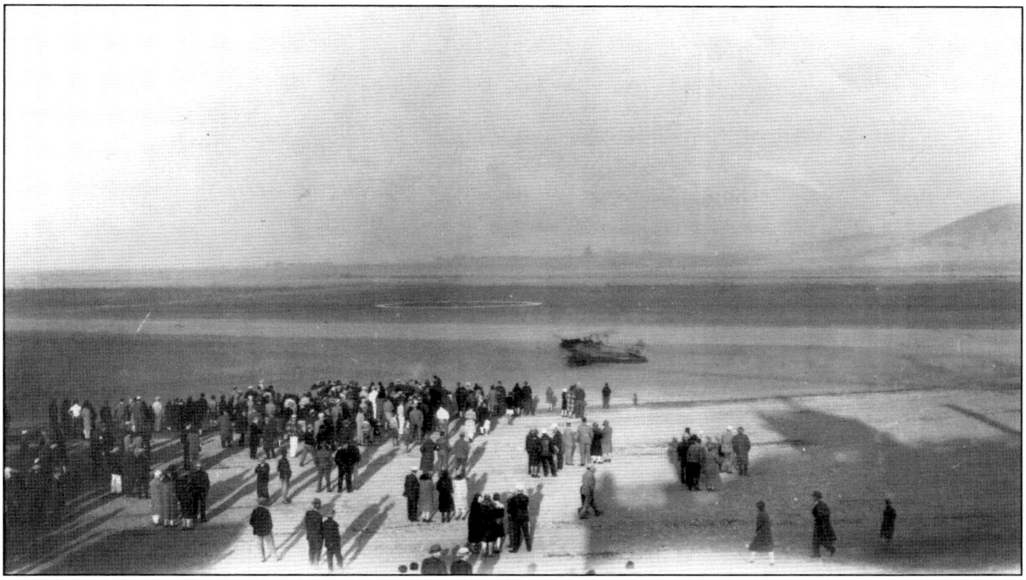

On December 1, 1929, a plane from the Copper King Flying Service and another from Hobi Airway Service of Eugene, piloted by Harold Sander, provided rides at the new Medford airport. The special rate for the day was one penny per pound. More than 400 people took advantage of the offer. (SOHS No. 9197.)

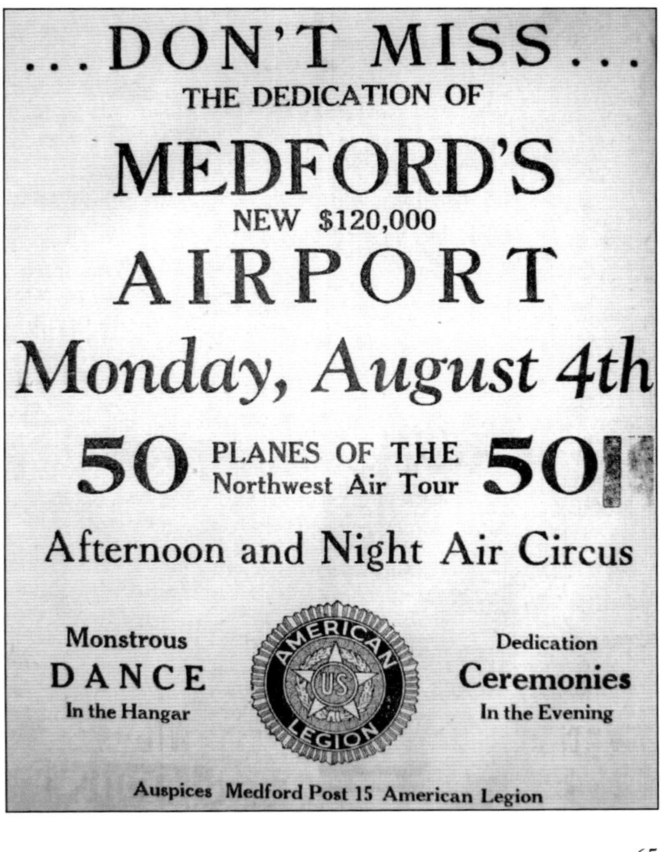

The official dedication of Medford's new $120,000 airport was a huge civic event, culminating in a "monstrous" dance in the new hangar. The dedication also coincided with the visit of the Northwest Air Tour, which provided aerial entertainment. (*Medford Mail Tribune*.)

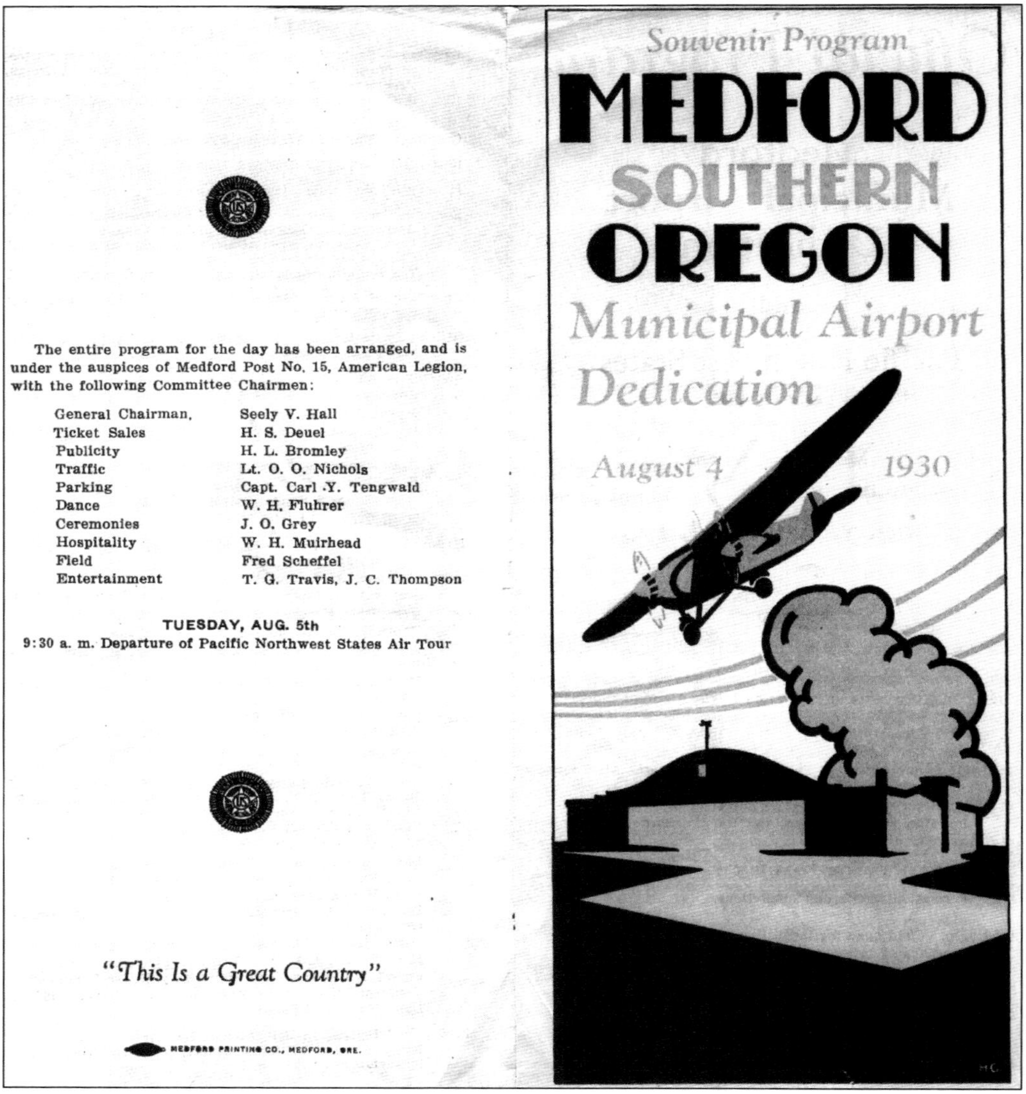

The entire program for the day has been arranged, and is under the auspices of Medford Post No. 15, American Legion, with the following Committee Chairmen:

General Chairman,	Seely V. Hall
Ticket Sales	H. S. Deuel
Publicity	H. L. Bromley
Traffic	Lt. O. O. Nichols
Parking	Capt. Carl Y. Tengwald
Dance	W. H. Fluhrer
Ceremonies	J. O. Grey
Hospitality	W. H. Muirhead
Field	Fred Scheffel
Entertainment	T. G. Travis, J. C. Thompson

TUESDAY, AUG. 5th

9:30 a. m. Departure of Pacific Northwest States Air Tour

"This Is a Great Country"

MEDFORD PRINTING CO., MEDFORD, ORE.

Souvenir Program

MEDFORD SOUTHERN OREGON

Municipal Airport Dedication

August 4 1930

The official dedication of the new Medford Municipal Airport was held on August 4, 1930. The festivities included the usual speeches, a visit and flying exhibition by the planes of the Northwest Air Tour, and a big dance in the expansive new hangar. (SOHS Manuscript.)

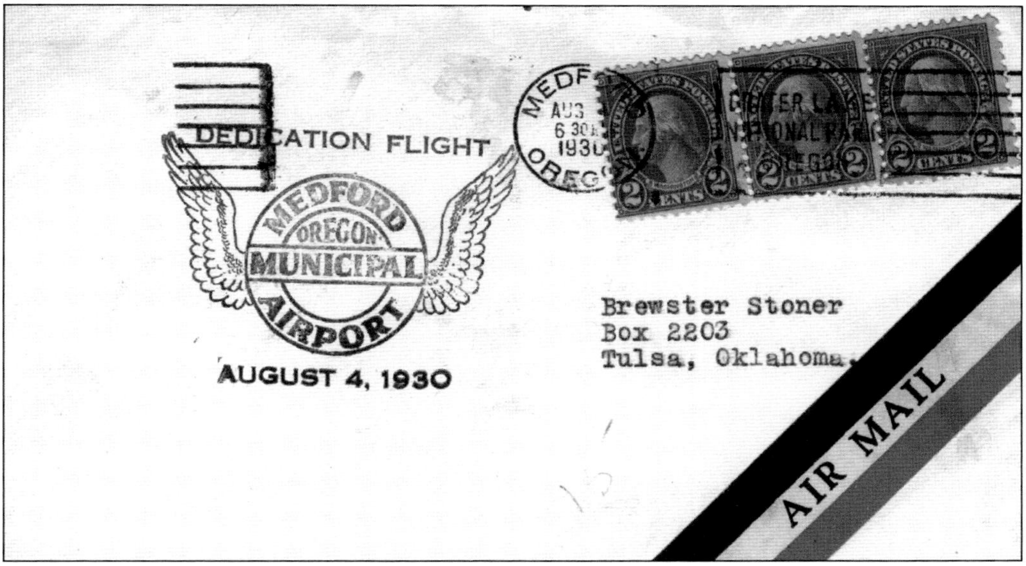

Also available for the dedication ceremonies were special dedication flight envelopes with a commemorative cachet stamp. These letters were carried on the airmail plane on dedication day. (SOHS Manuscript.)

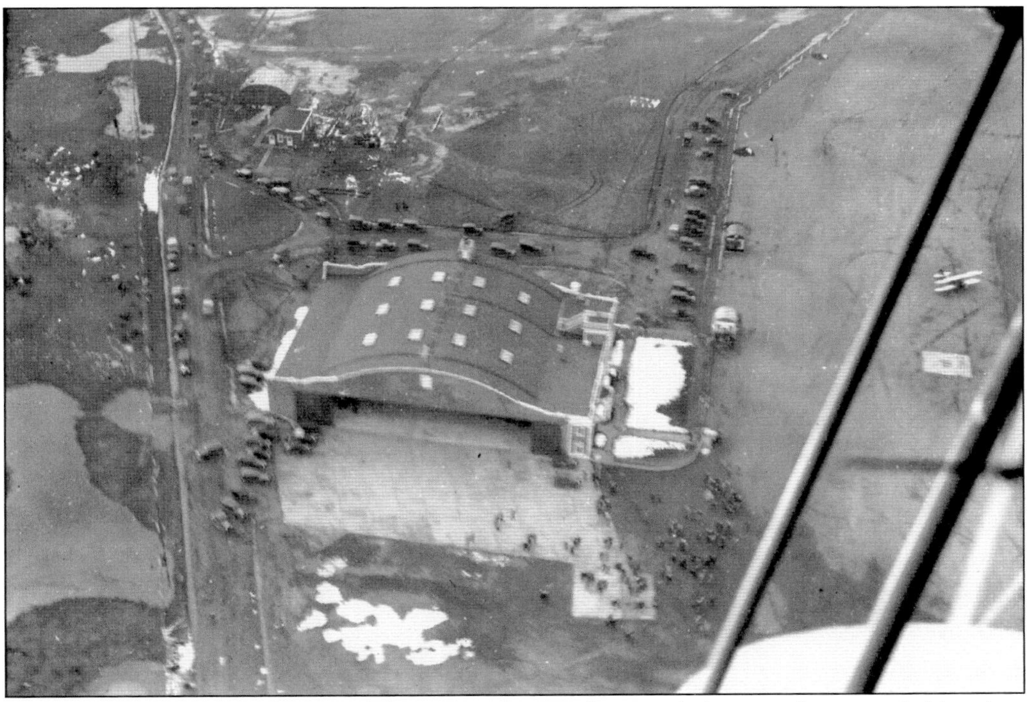

Judging by the number of cars and the traces of snow, this aerial photograph was probably taken in December 1929, when "penny a pound" rides were being given. This view shows just how expansive the new hangar was in relation to the attached administrative building. (Rogue Valley International-Medford Airport.)

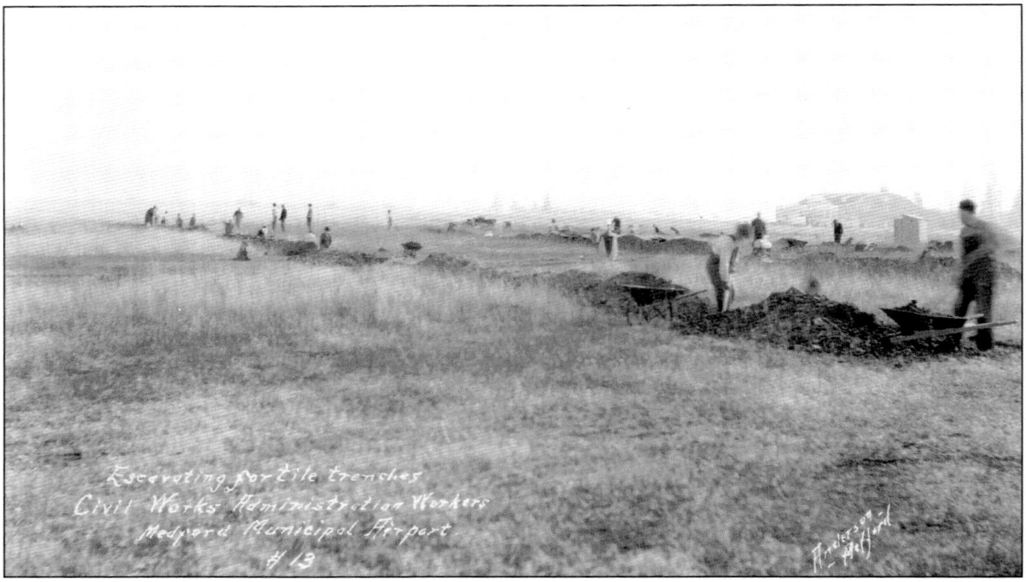

Great care was taken to ensure proper drainage of the Medford Municipal Airport runways. The Civil Works Administration hired local men in need of employment to dig additional trenches that would be lined with tiles between the fall of 1933 and the spring of 1934, the period when the CWA was in operation. (Rogue Valley International-Medford Airport.)

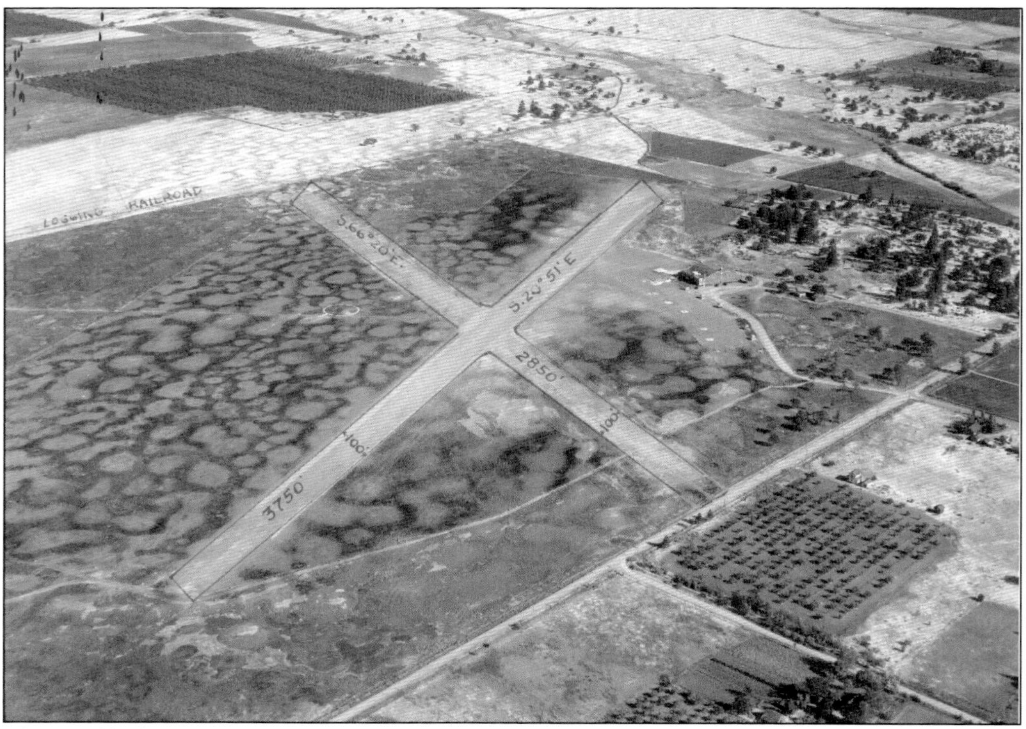

The Medford Municipal Airport was constantly being improved, expanded, and updated over the years. This aerial photograph from 1935 shows that a cross runway was added to the original. (Rogue Valley International-Medford Airport.)

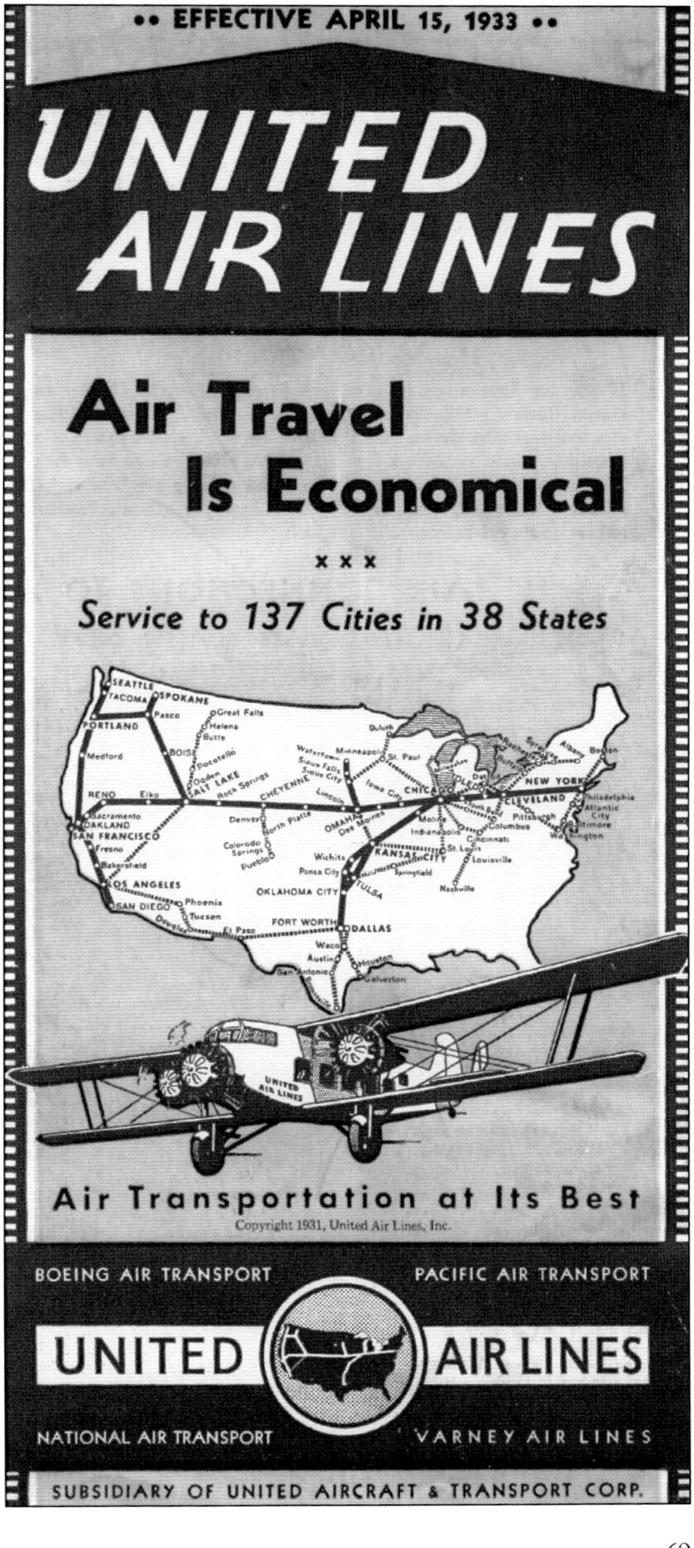

Throughout the 1930s, activity at the Medford Municipal Airport slipped into a regular routine. The merger of Boeing Air Transport, Pacific Air Transport, National Air Transport, and Varney Airlines was reorganized as a new company, United Airlines. United Airlines continued the regularly scheduled service it had provided since the arrival of Pacific Air Transport. (SOHS Manuscript.)

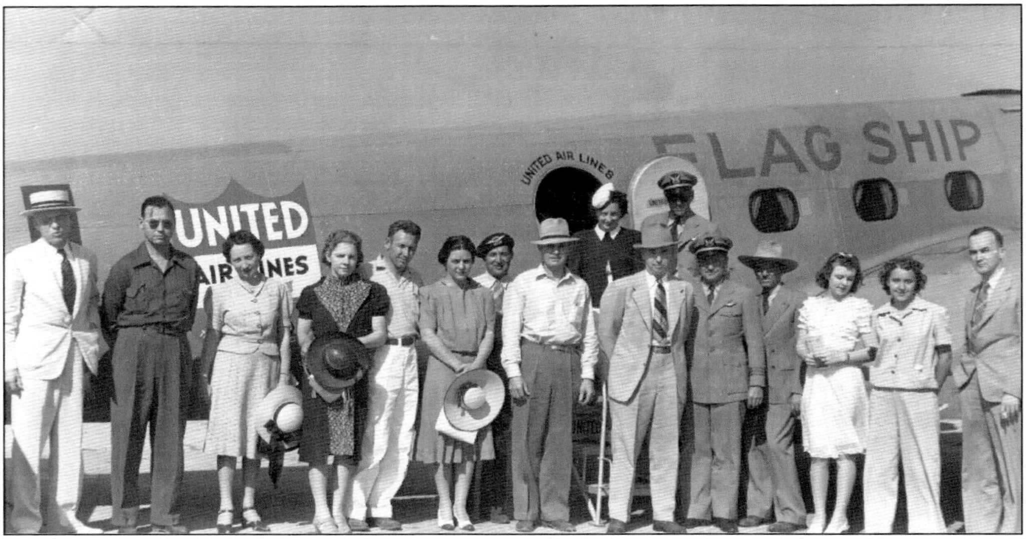

In 1933, Boeing introduced the Model 247, considered by most to be the first modern airliner. Most of the initial production run was sold to United Airlines, but the plane's popularity quickly waned with the introduction of the larger and more capable DC-3. In this image, passengers and crew pose alongside a Boeing 247 bearing the 1936 United Airlines logo. (SOHS No. 16791.)

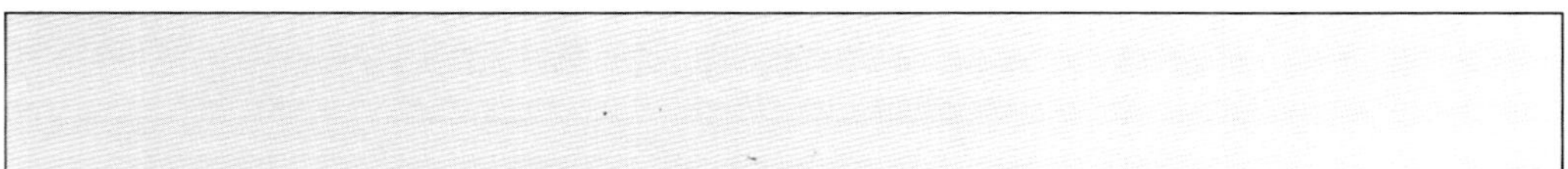

A United Airlines DC-3 taxis toward the terminal at the Medford Municipal Airport. United began acquiring DC-3s in the fall of 1936 to replace their less successful Model 247. The Douglas DC-3 revolutionized the air transport system. Its profitability freed many carriers from their reliance on government mail subsidies. (SOHS No. 19200.)

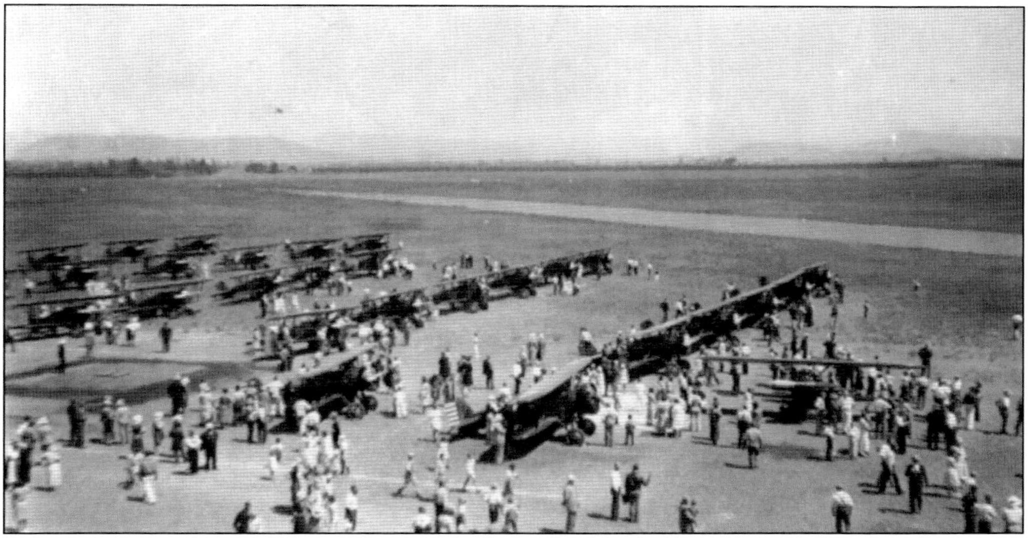

A visiting squadron of aircraft can always be counted on to draw visitors to any airfield, and Medford was no exception. These Boeing P-12 pursuit planes generated a great deal of interest in this image, taken from the top of the administration building at the airport on June 12, 1931. (Rogue Valley International-Medford Airport.)

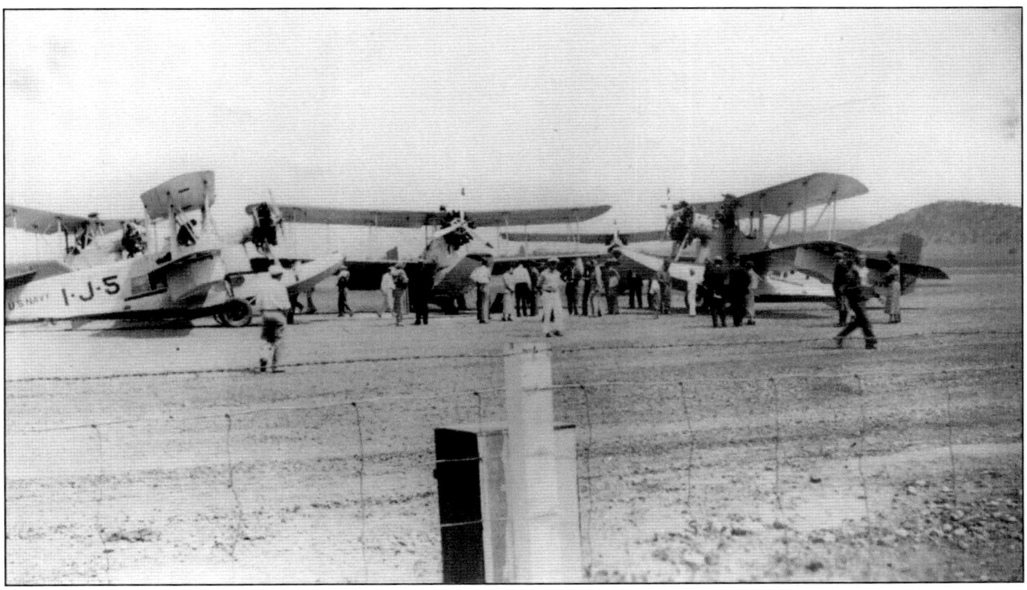

Medford was a frequent stop for military aircraft travelling up and down the West Coast. This photograph, taken around 1930, shows a U.S. Navy Loening OL-8 amphibian aircraft that made a stop at the Medford airport. (SOHS No. 12731.)

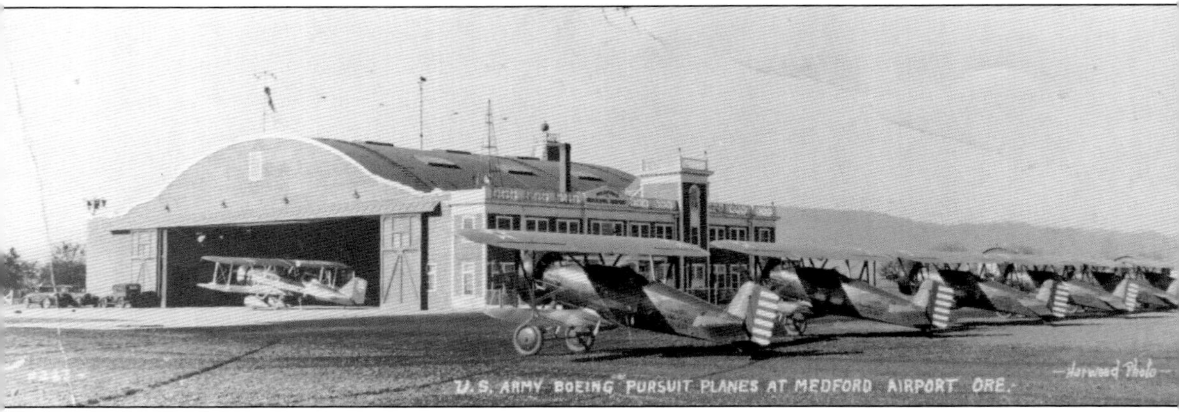

Six Boeing P-12 Ds are parked outside the Medford terminal around 1930. The nimble little P-12s served as the mainstay of the U.S. Army Air Corps's pursuit squadrons, and the D models were the last fabric-covered fighters. Later models had an aluminum-covered fuselage. These were the last biplane pursuit aircraft used by the U.S. Army. Parked in front of the hangar door is a Boeing Model 40-B mail plane. (SOHS No. 17041.)

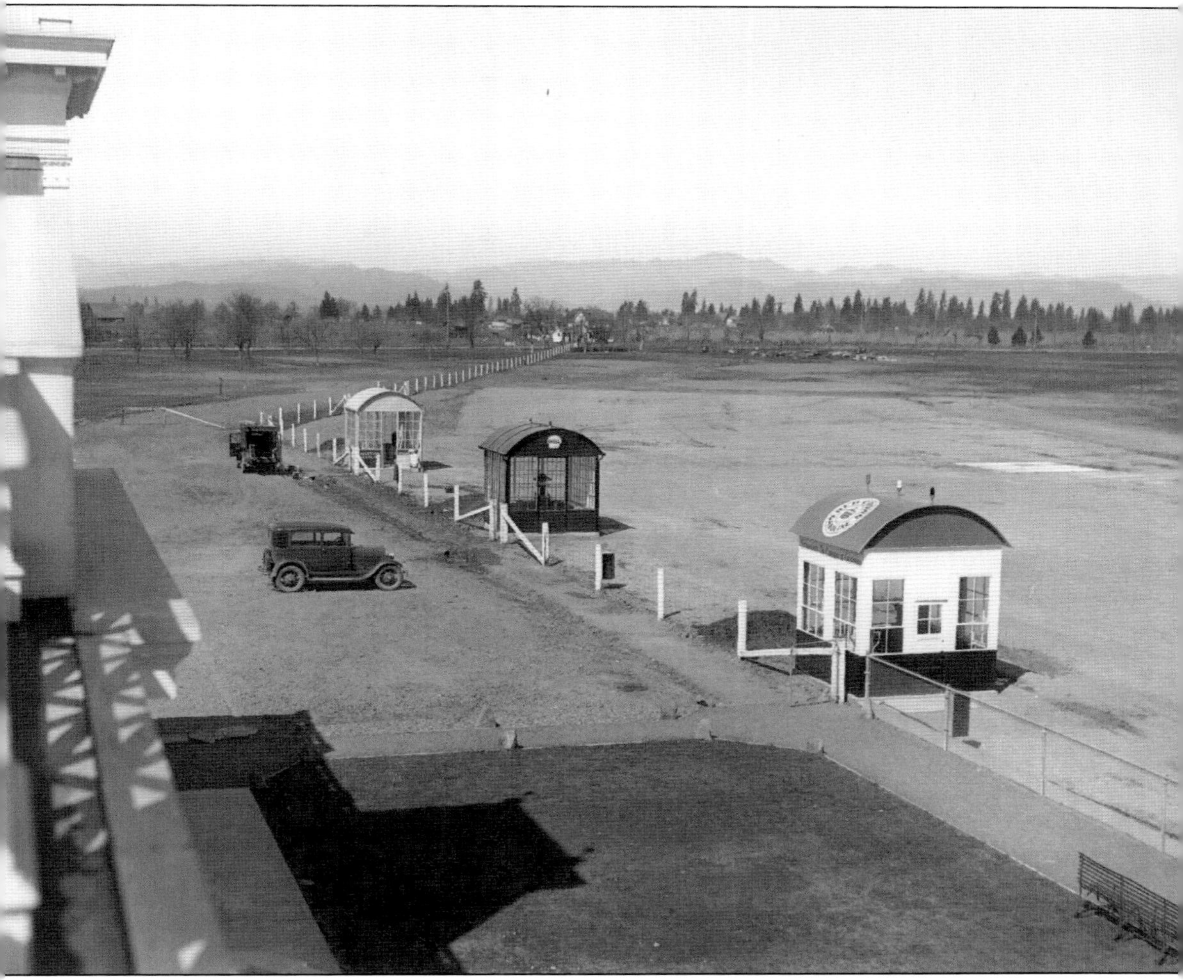

The Medford Municipal Airport was continually being improved upon. This image, taken from the terminal roof on April 13, 1934, shows the expanding variety of aviation fuel available: Red Crown, Shell, and Richfield. The ready availability of fuel helped ensure Medford's popularity as a stop for transient civil and military aircraft. (SOHS No. 19186.)

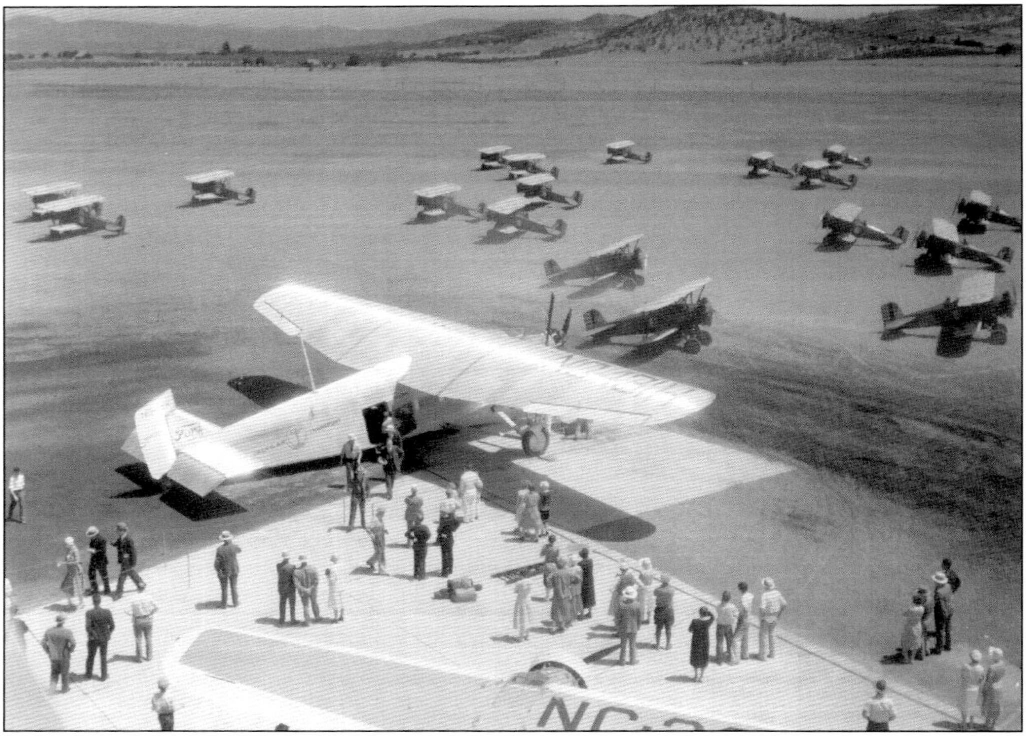

Passengers disembark from a Ford trimotor operated by Pacific Air Transport in 1931. Also visible on the field is a visiting squadron of Army Boeing P-12 pursuit planes. Military aircraft were regular visitors to Medford when on maneuvers or in transit. (SOHS No. 960.)

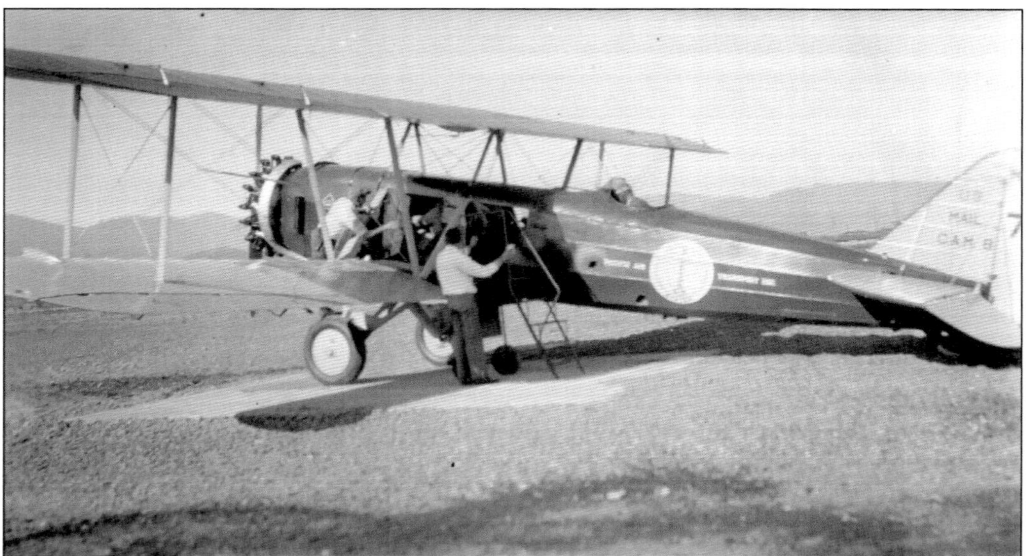

Boeing Aircraft introduced the Model 40-B passenger plane to its fleet in 1928. It was designed to accommodate mail and four passengers, who rode in the relative comfort of two small cabins. (SOHS No. 956.)

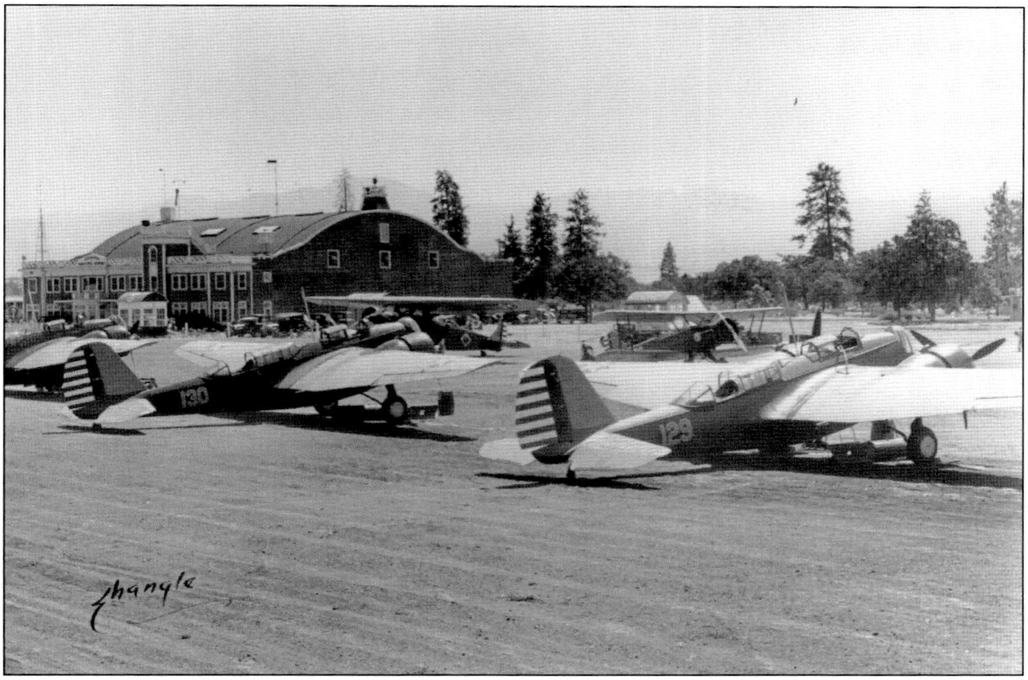

The U.S. Army Air Corps recognized the public relations advantages of having its aircraft on tour, and a public open house was standard. Martin B-10 bombers set up a temporary base of operations at Medford in the summer of 1935, drawing a large crowd. (SOHS No. 20714.)

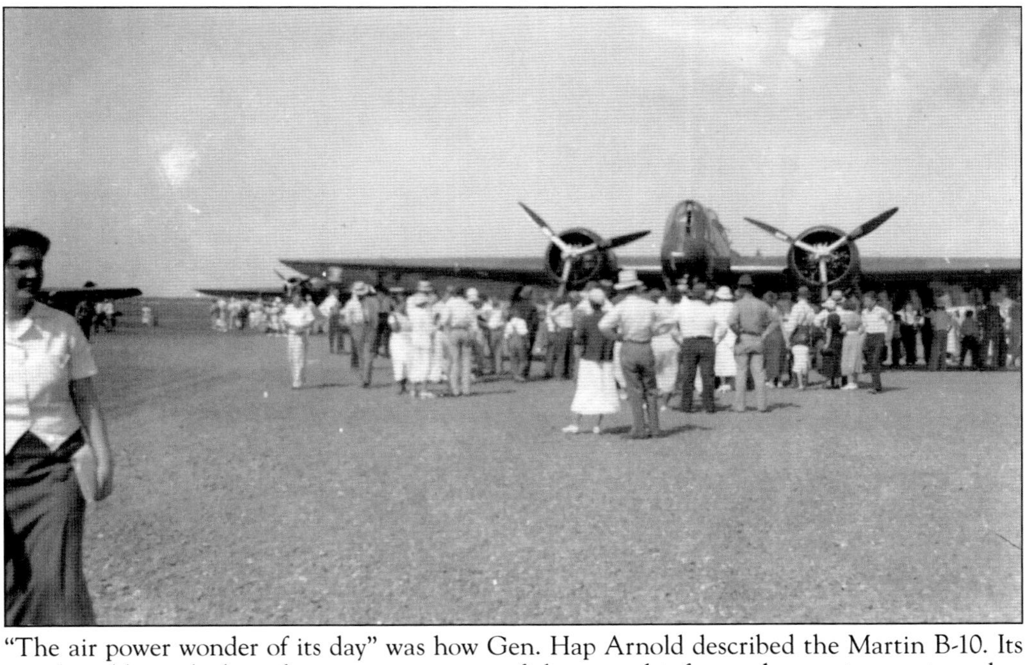

"The air power wonder of its day" was how Gen. Hap Arnold described the Martin B-10. Its speed could match that of many contemporary fighters, and it featured many innovations that were incorporated into its successors. The 18 bombers that visited Medford attracted interested crowds. (SOHS No. 20710.)

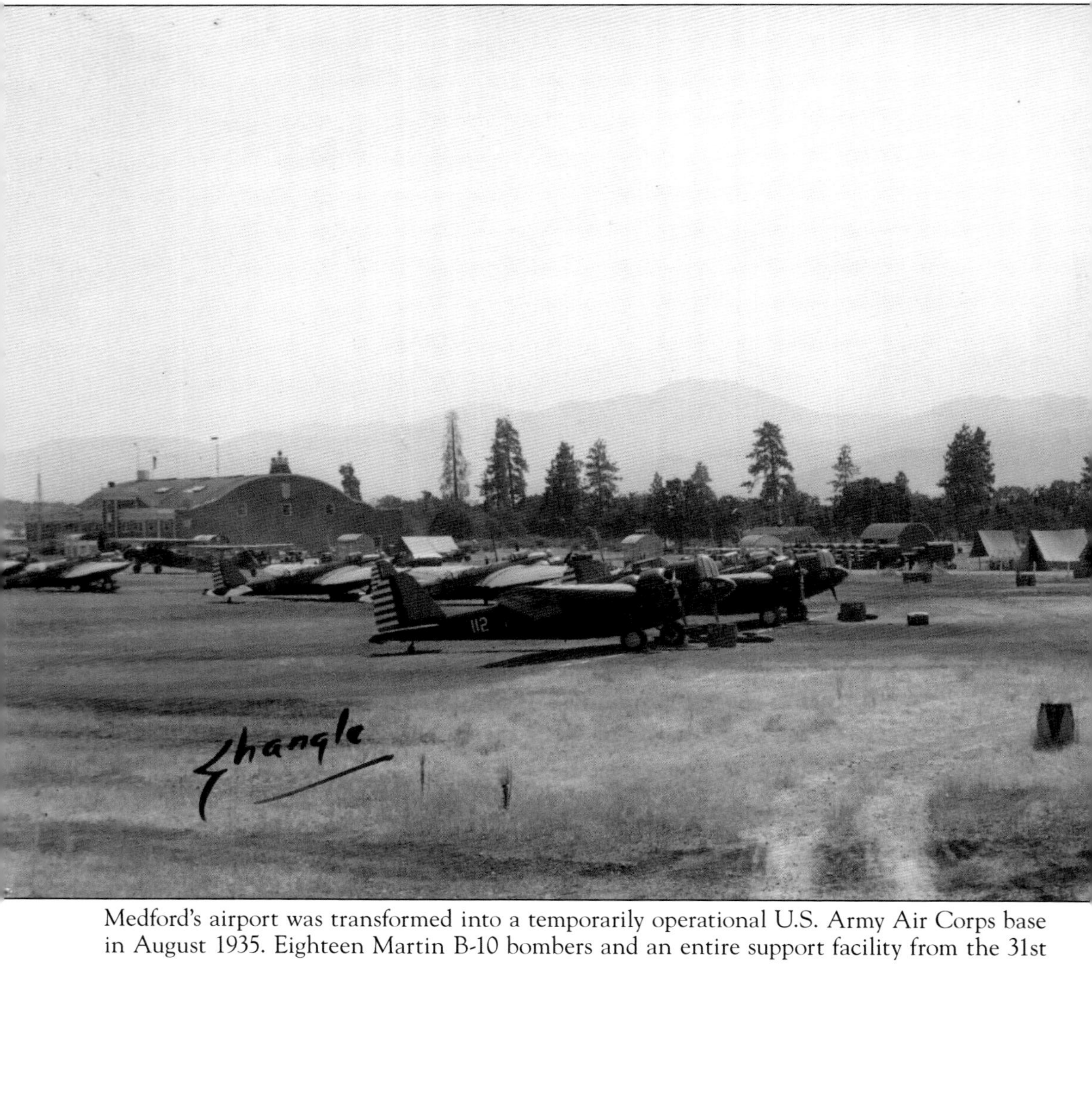

Medford's airport was transformed into a temporarily operational U.S. Army Air Corps base in August 1935. Eighteen Martin B-10 bombers and an entire support facility from the 31st

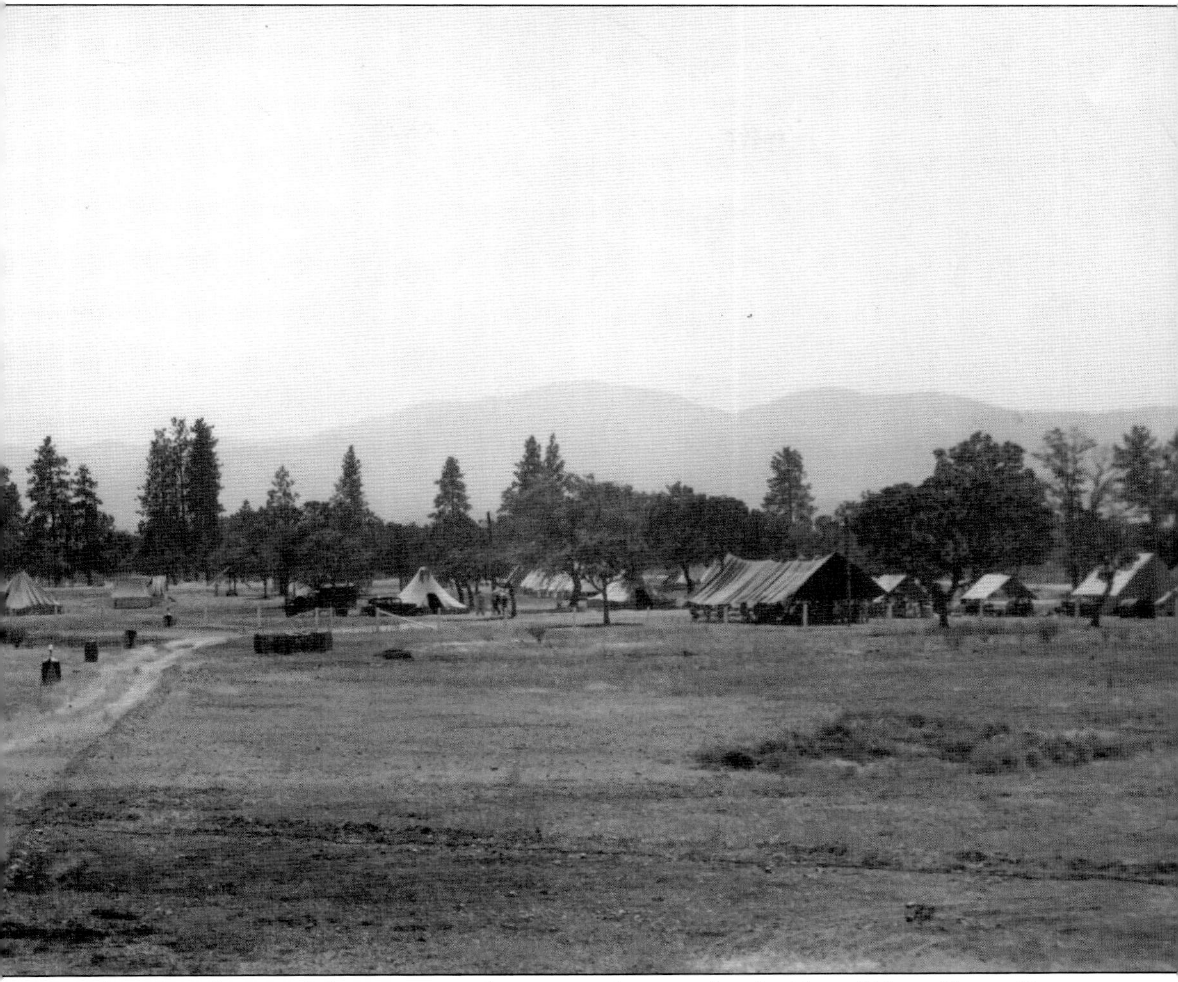

Bombing Squadron used the airport as a base to conduct simulated raids on Salem, Grants Pass, and Dunsmuir. (SOHS No. 6916.)

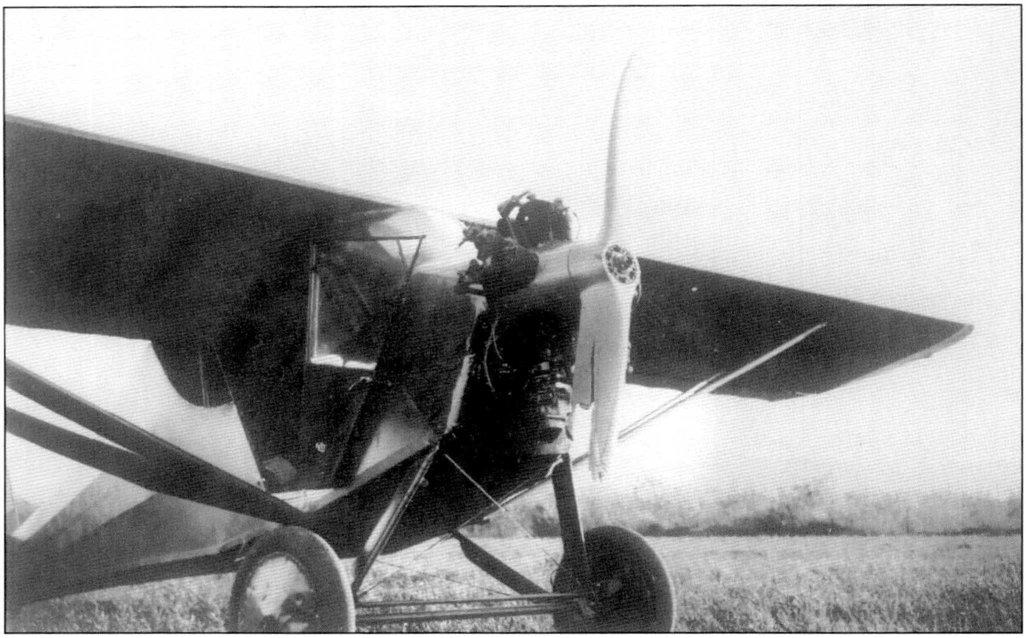

The Velie Monocoupe was introduced in 1927 and advertised as "the ultimate plane for the private flyer." There are obvious signs of a mishap with this particular Monocoupe. While being flown by Aubrey Sander, one of the five cylinders blew on the 62-horsepower engine, destroying the propeller. Sander was unable to return to the airport but made a safe landing in a field near Dakota and Stewart Streets. (SOHS No. 12728.)

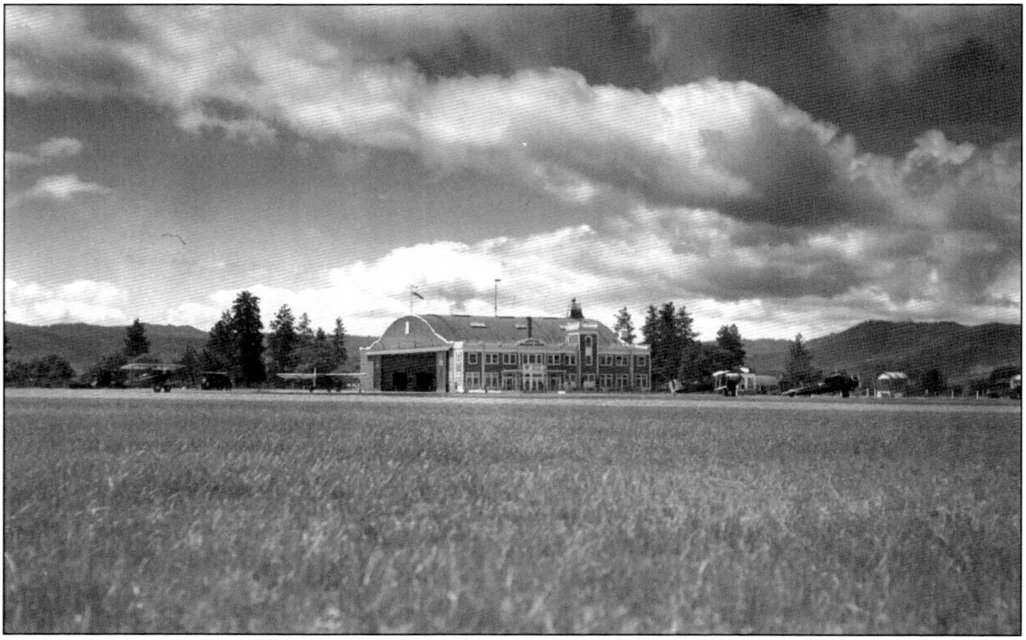

Tranquil settings like the one pictured here were not to last. After the Japanese attack on Pearl Harbor, much changed at the Medford Municipal Airport. The entire facility was leased to the U.S. Army as it prepared to do its part in the war. (SOHS.)

Five

The Airport Goes to War

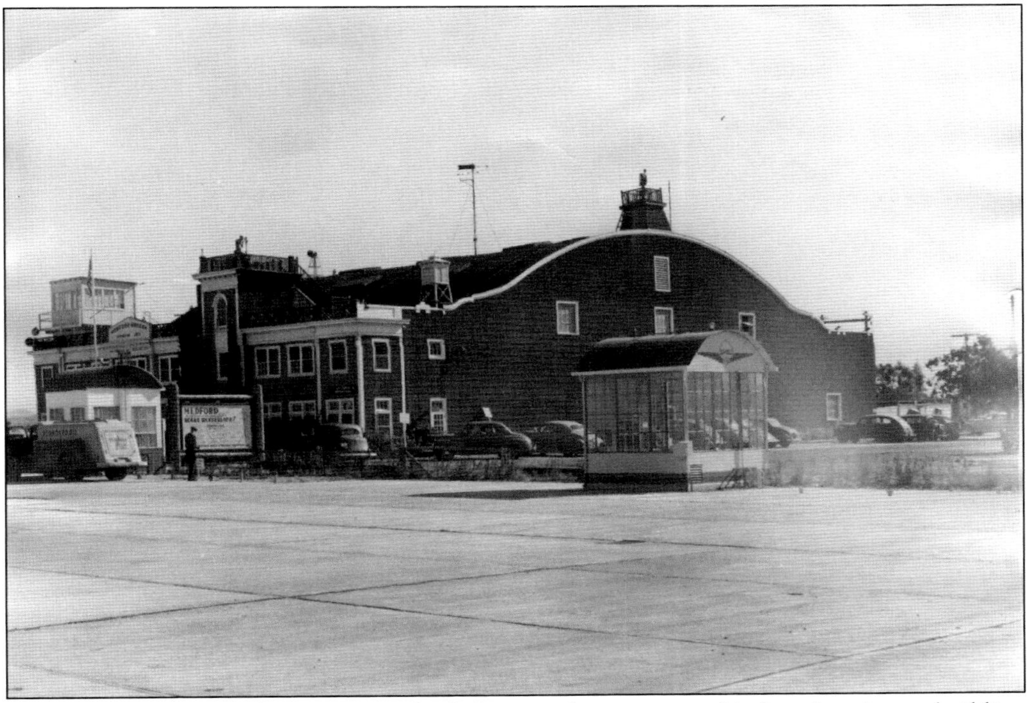

While under lease to the army, a makeshift control tower was added to the airport building on the corner nearest the runway. With commercial traffic relocated to the nearby United Airlines building, however, few visitors were able to appreciate the "Heart of the Beautiful Rogue Wonderland" welcome sign. (SOHS No. 18017.)

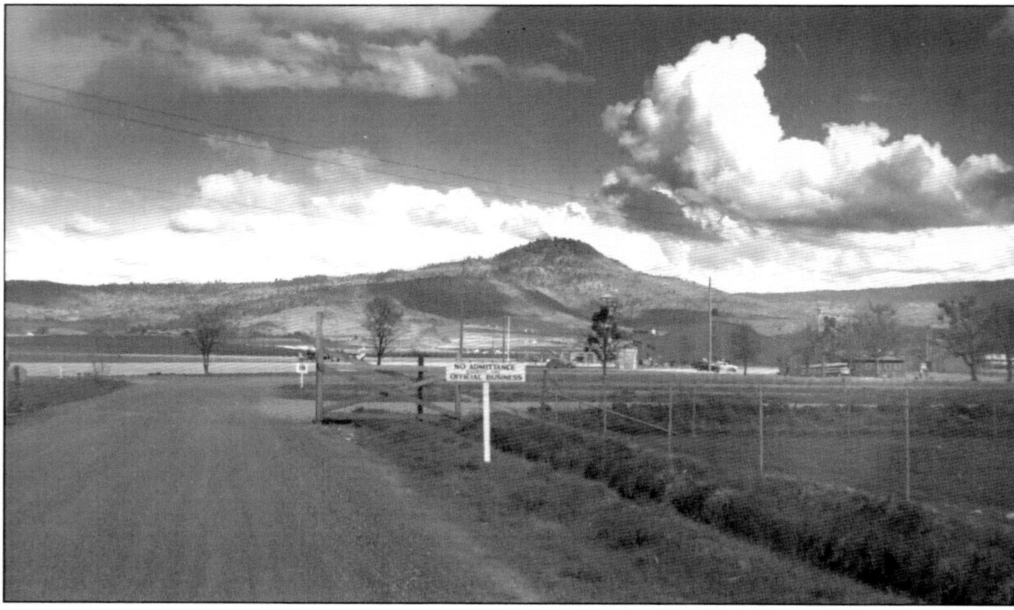

In early 1942, according to the *Medford Mail Tribune*, the U.S. Army leased the Medford airport for the "duration of the war plus six months." With the exception of the regularly scheduled United Airlines flights, the Medford Municipal Airport became off limits to unauthorized civilian personnel; as the erected sign states, "No Admittance Except On Official Business." (SOHS No. 20733.)

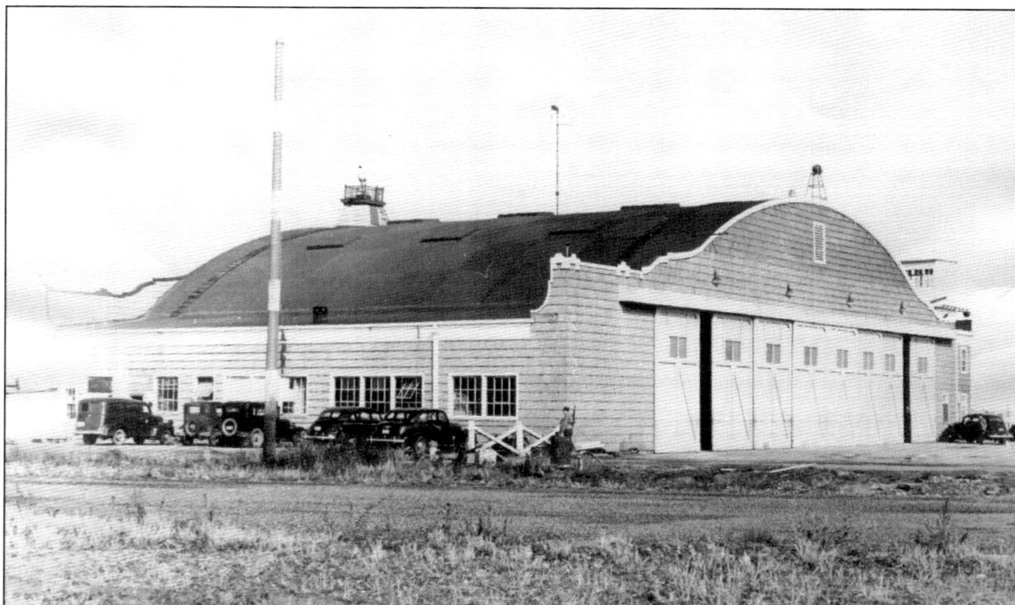

The U.S. Army leased the Medford Municipal Airport from the city for $1 per year during the Second World War. The army used the facility to service military air traffic traveling up and down the West Coast. In this image, a sentry can be seen at the corner of the hangar. (SOHS No. 20704.)

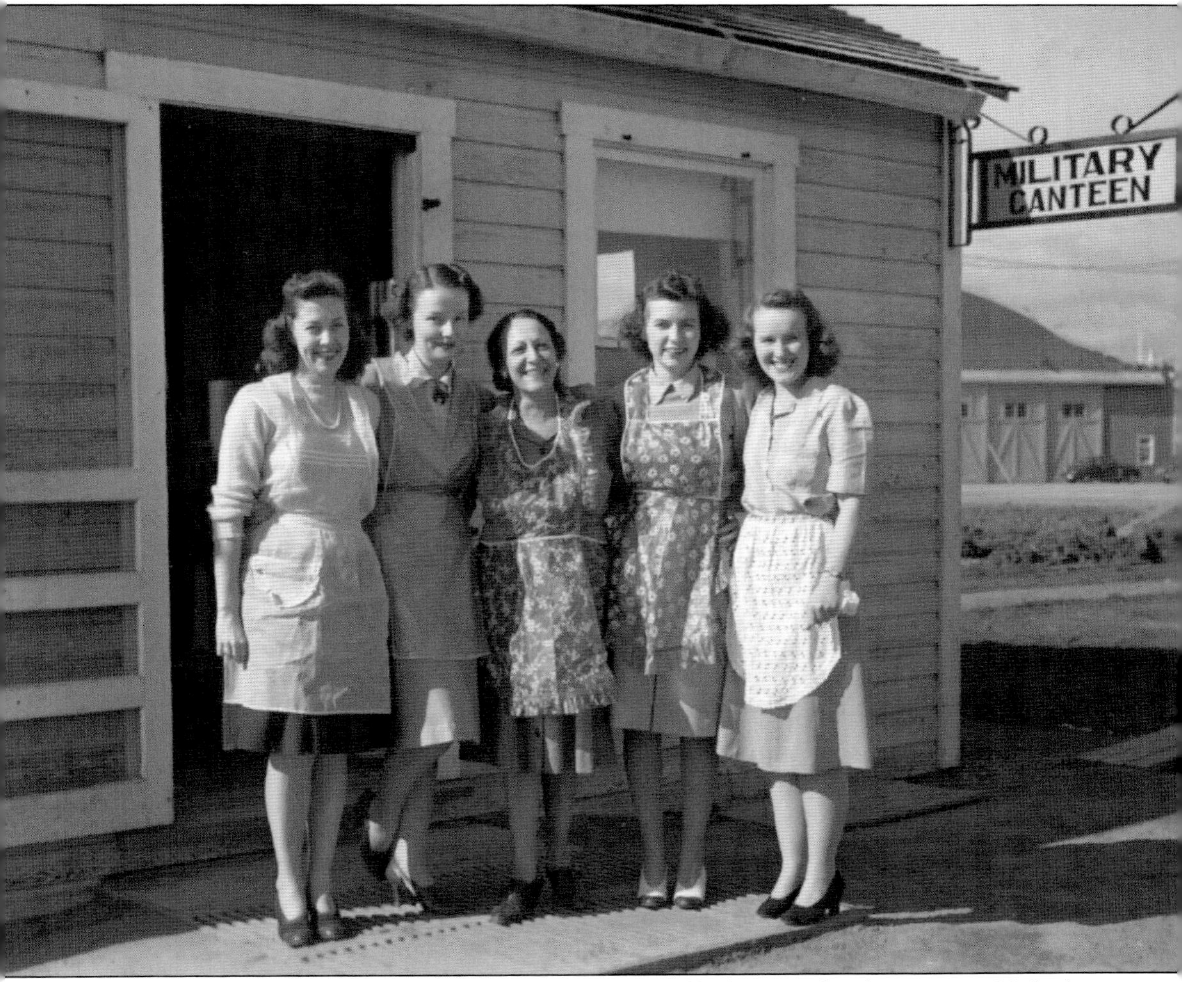

When the U.S. Army Air Corps assumed operation of the Medford airport, local citizens established the Wing In Canteen to provide food and relaxation for soldiers and aircrews passing through. Pictured here are, from left to right, Peachie Drury, Isabel Miller, former Broadway ingénue Grace Fiero, Kay Conroy, and Pat Thompson, the author's mother. By the end of the war, the kitchen was serving an average of 400 meals each day. (Author's collection.)

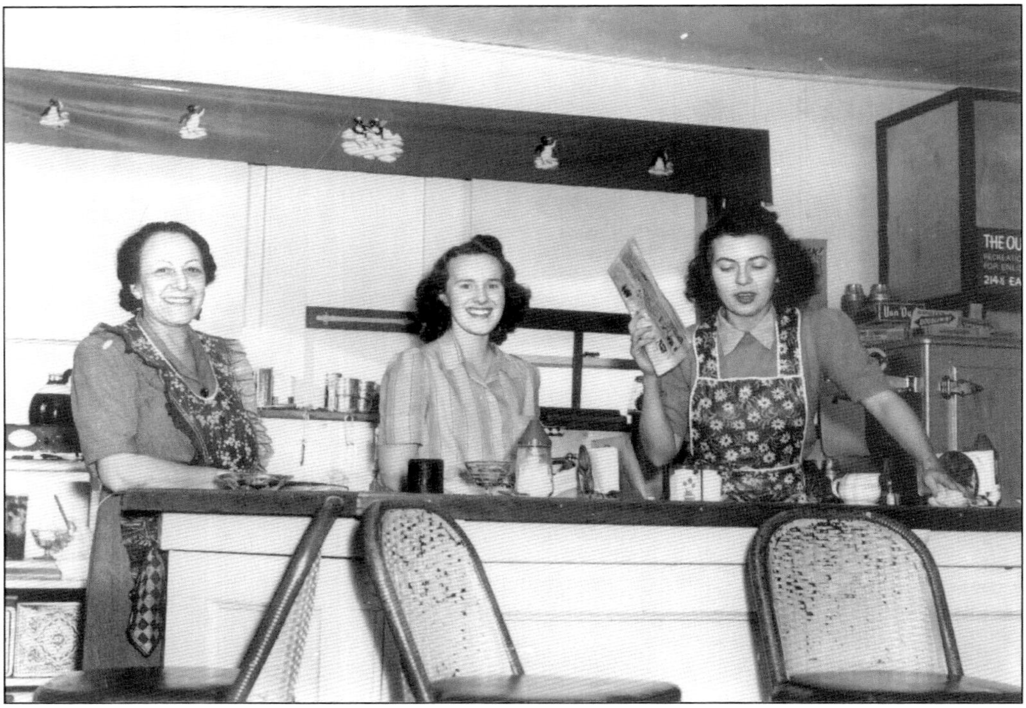

From left to right, Grace Fiero, Pat Thompson, and Pat Conroy were regular volunteers at the Medford's Wing In Canteen during the war. A close examination reveals that the counter seats are the same ones that once graced the café counter at Medford's original airport when it opened in 1929. (SOHS No. 16809.)

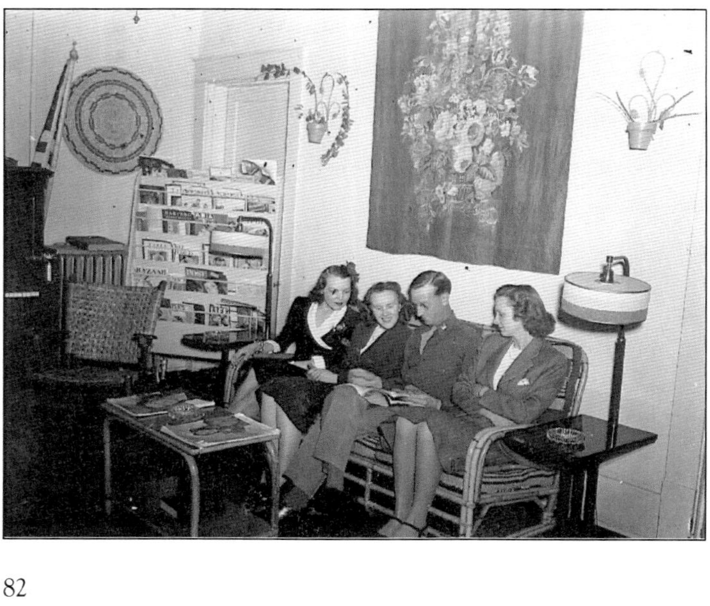

In addition to the hamburgers and coffee served at the counter, the Wing In Canteen provided a place to relax, read, write letters, and talk with the girls. After the canteen was expanded, a piano was brought in, visible on the left. One general was quoted as praising the Wing In Canteen as one of the best airport canteens in the country, saying it "was known the world over." (Author's collection.)

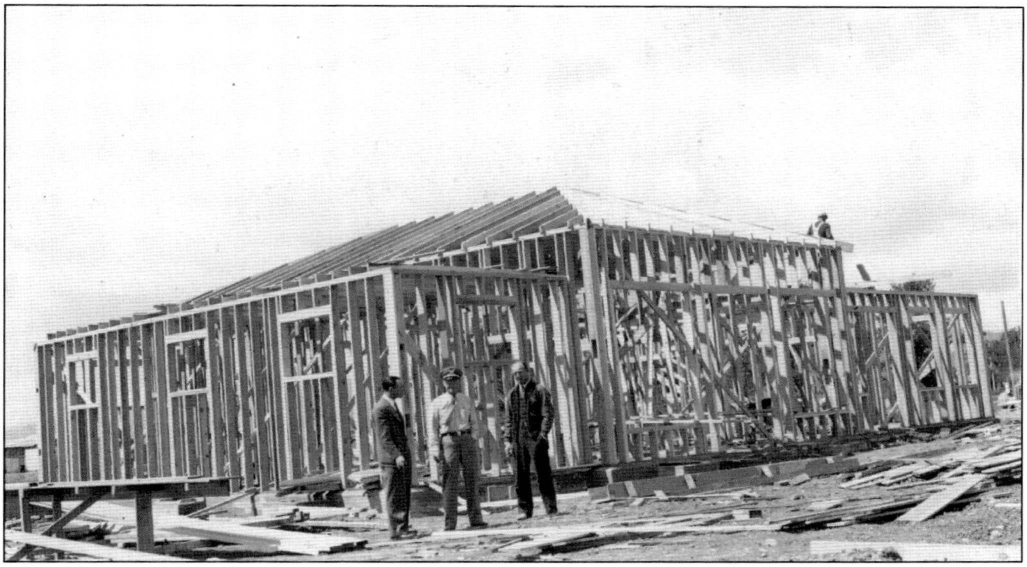

With the army occupying the airport terminal and administrative building, it became necessary to construct a new passenger and ticketing facility for United Airlines, which maintained its flying schedule during the war. (SOHS No. 20274.)

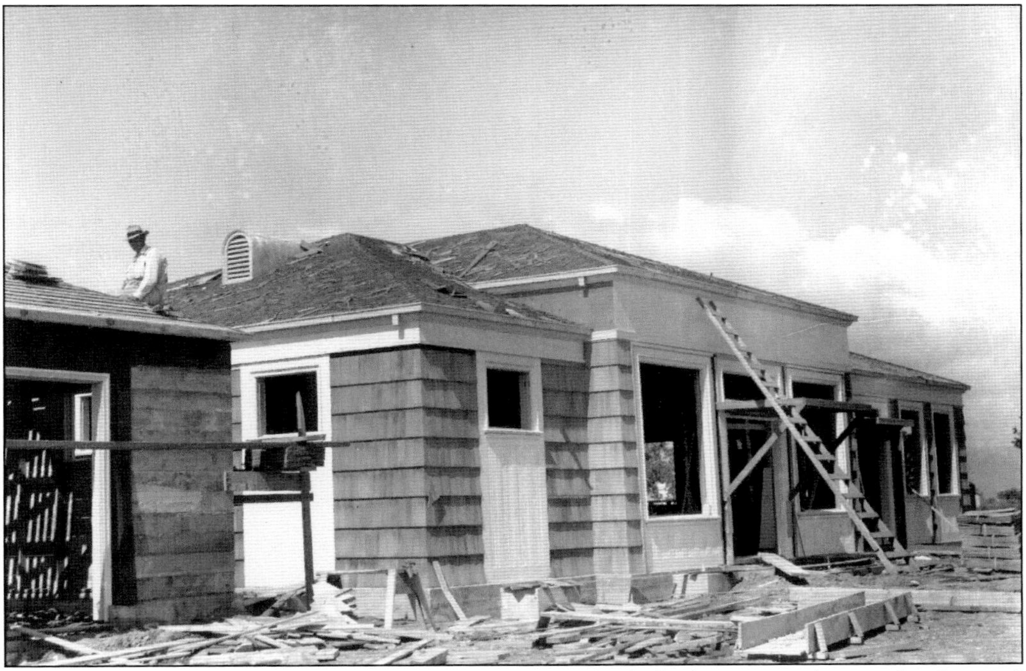

The new United Airlines terminal was designed by Frank Clark, the same architect who designed the main airport building. His new structure was created to compliment the earlier structure. (SOHS No. 20263.)

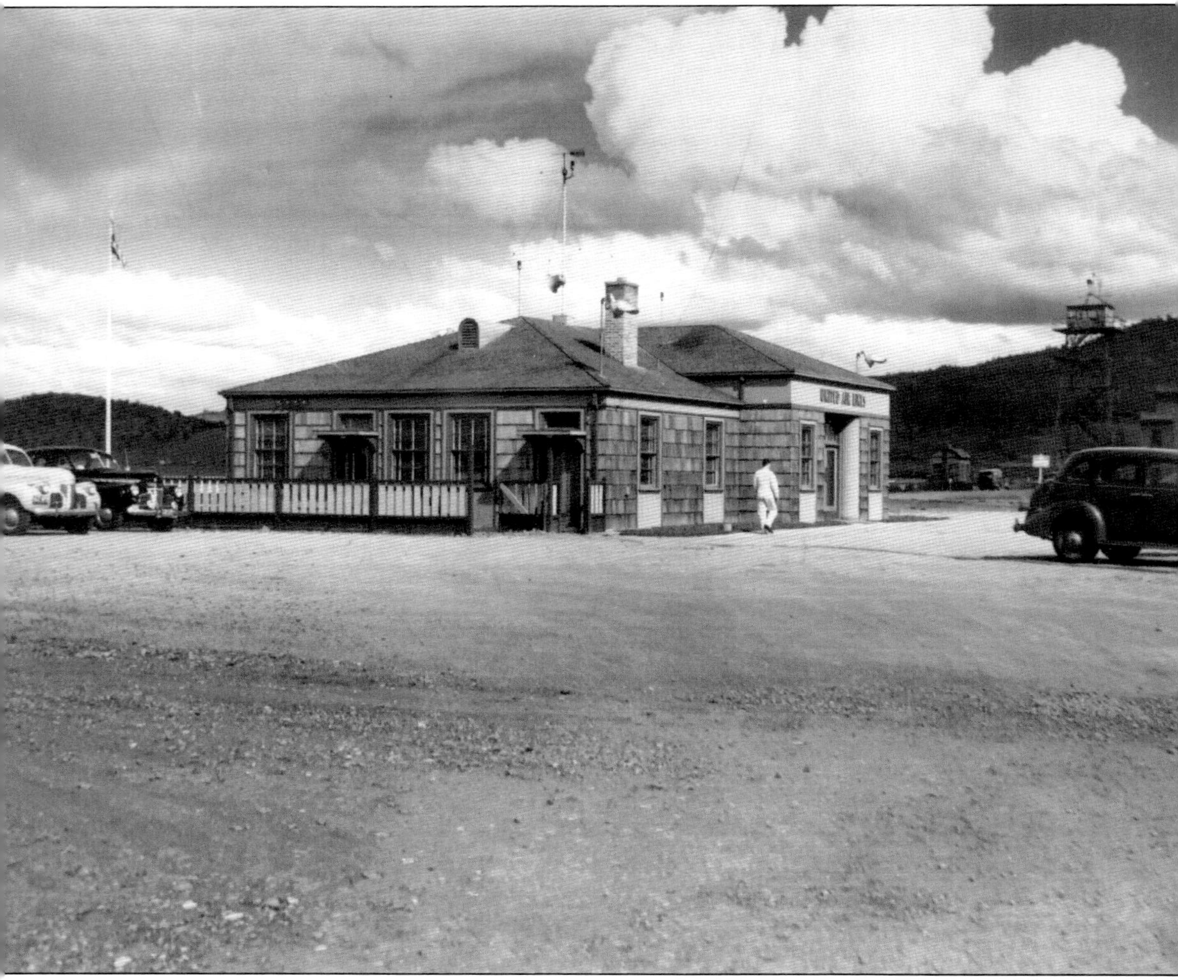

The new United Airlines terminal building created a separate section for commercial air traffic at the Medford Municipal Airport. What little civilian flying was allowed during the war was moved back to the original Newell Barber Field. In the distance is a new free-standing control tower, augmenting the earlier one built on the roof of the administration building. (SOHS No. 3725.)

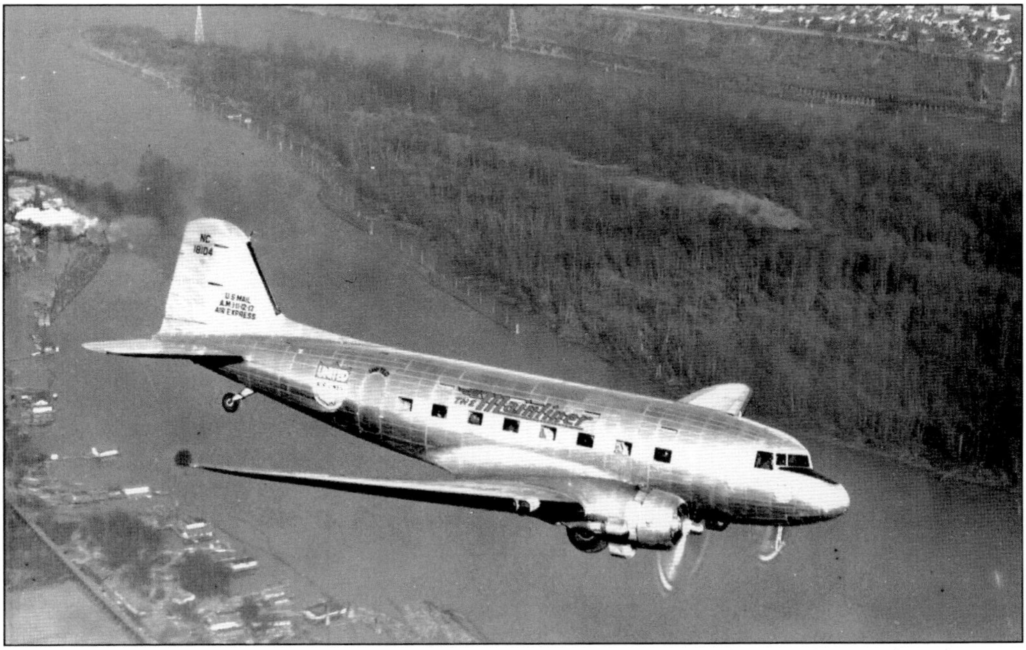

From its introduction in the mid-1930s until the development of larger planes after the war, the Douglas DC-3 was the workhorse of commercial aviation. A United Airlines DC-3 departs Portland's Swan Island airport for Medford, the next stop on the West Coast route. (Pearson Air Museum.)

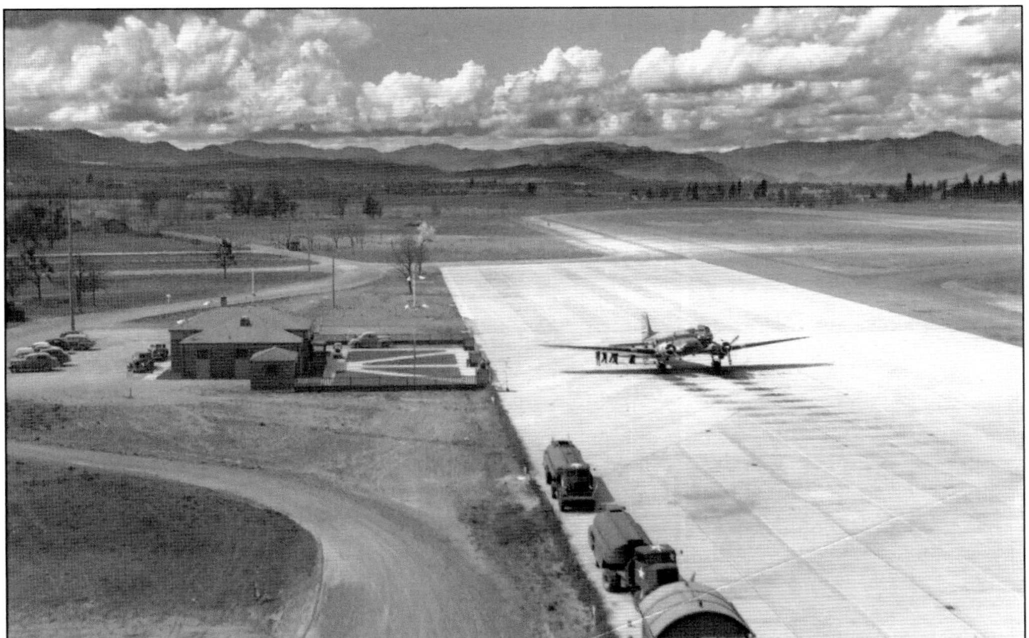

Passengers prepare to board a United Airlines DC-3. The presence of army fuel trucks suggests that this photograph was taken during World War II from the top of the administration building or the adjacent tower. (SOHS No. 16667.)

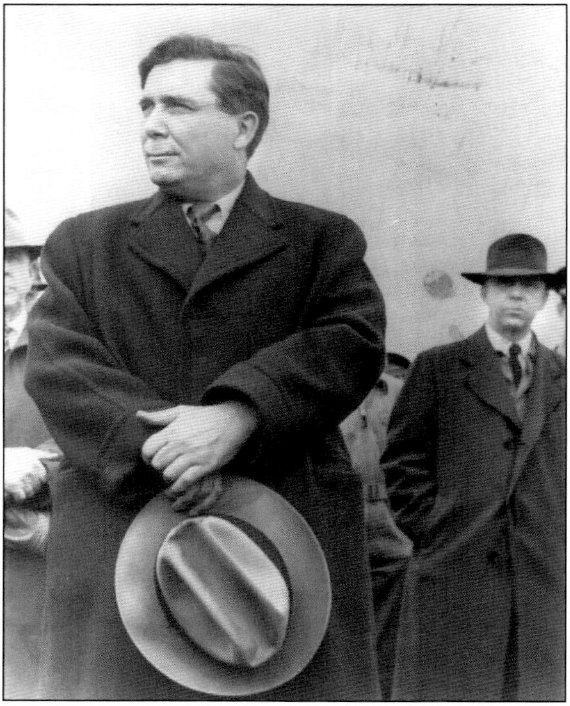

Political candidates soon discovered that airplanes were much more efficient than trains for "whistle stop" campaigning; more ground could be readily covered in less time. On February 14, 1944, Wendell Willkie made an appearance at the Medford airport in his short-lived second attempt to win the nomination for president. Soon after his visit, he withdrew from the race and died shortly thereafter. Behind Willkie is Medford resident Carl Tengwald. (SOHS No. 15224.)

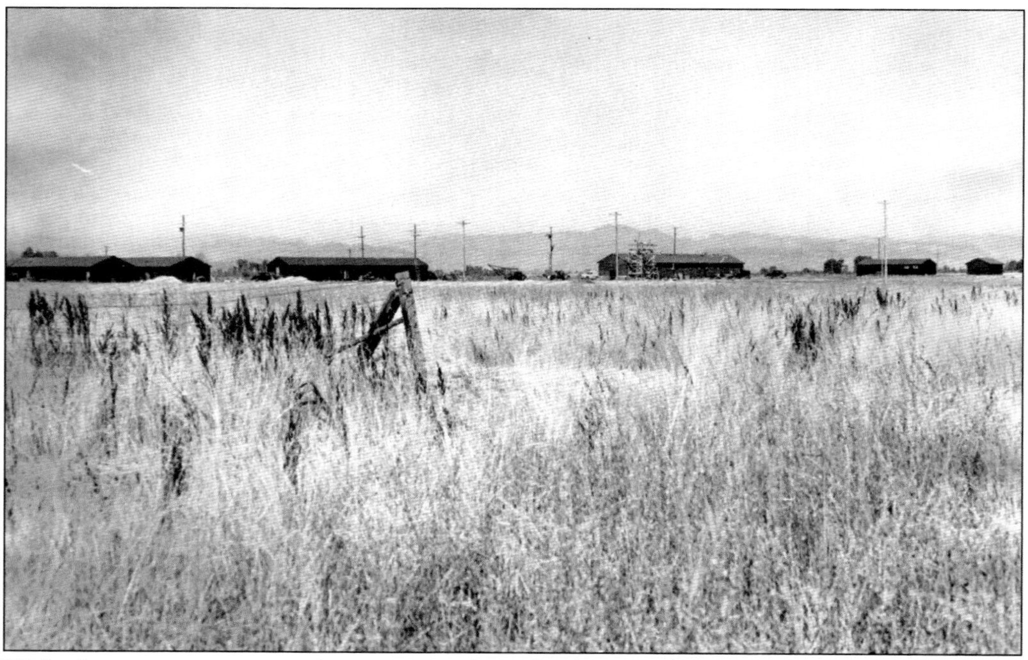

With the army presence at the airport and nearby Camp White steadily growing as the war progressed, it became inevitable that the army would expand the size of the airport and build numerous temporary buildings. The purchase of additional land around the airfield ultimately complicated and delayed the return of the airport to the city at the conclusion of the war. (SOHS No. 20743.)

Six

After the War

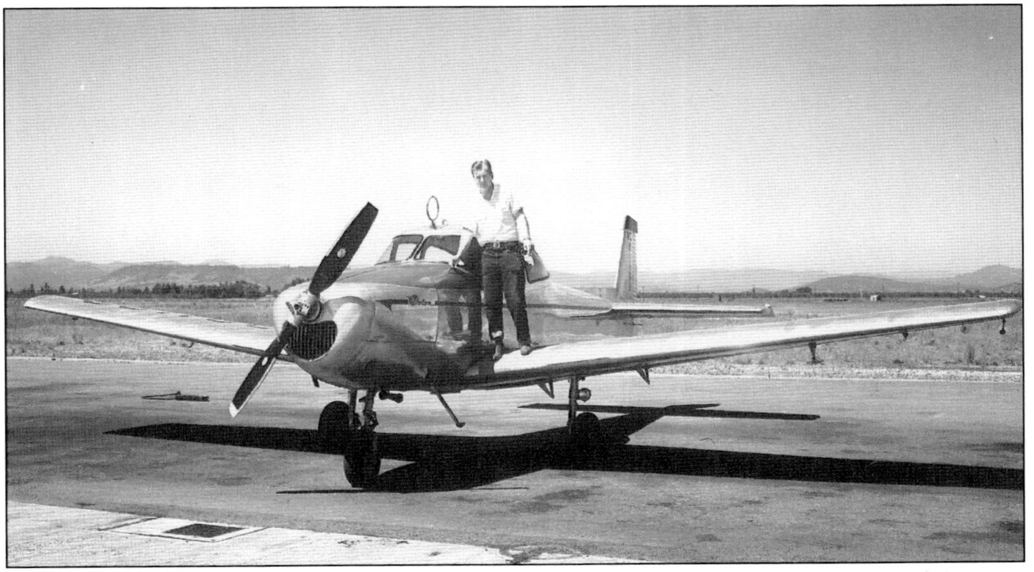

The end of World War II brought the return of civil aviation to the Medford Municipal Airport, and activity at the field increased dramatically. One rapidly expanding group was dubbed "the Flying Lumbermen" by the local newspaper. They were men who began using their own aircraft in the course of their business and leisure. One of these men was Dick Alley, seen here in 1946 with his new North American Navion, the first of these popular aircraft delivered to Oregon. (Author's collection.)

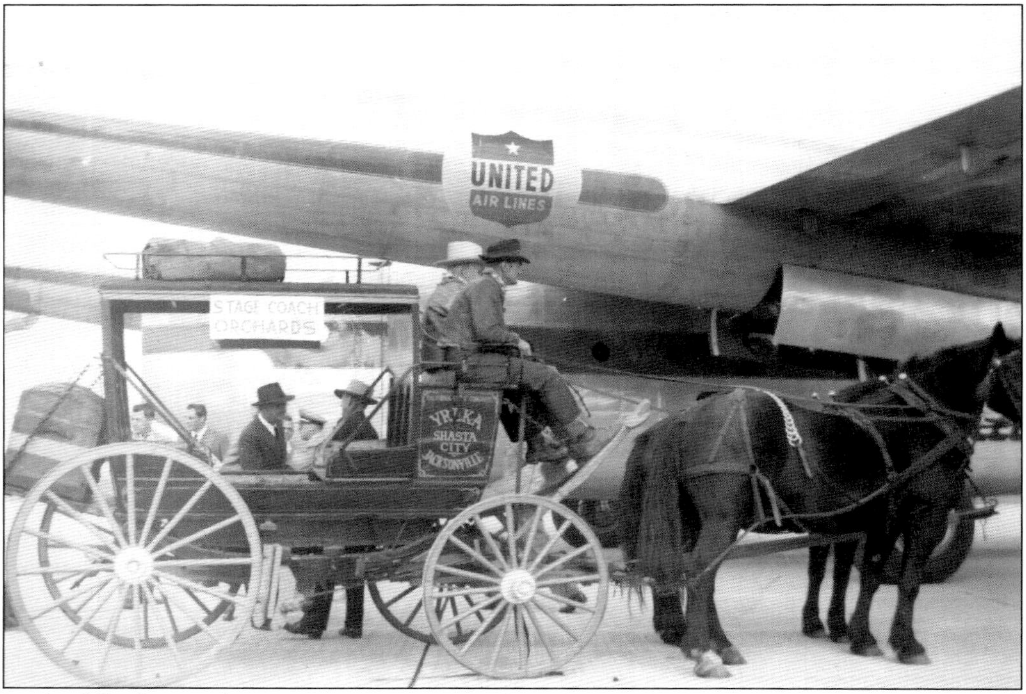

In the fall of 1946, the U.S. Army Air Corps leased a Fairchild C-82 to United Airlines. The plane, dubbed the "Flying Mail Car," had been reconfigured to act as a full postal sorting station, much like those long-used by the railroads. After less than two weeks, however, the experiment was scrapped and the plane returned to military service. (SOHS No. 16789.)

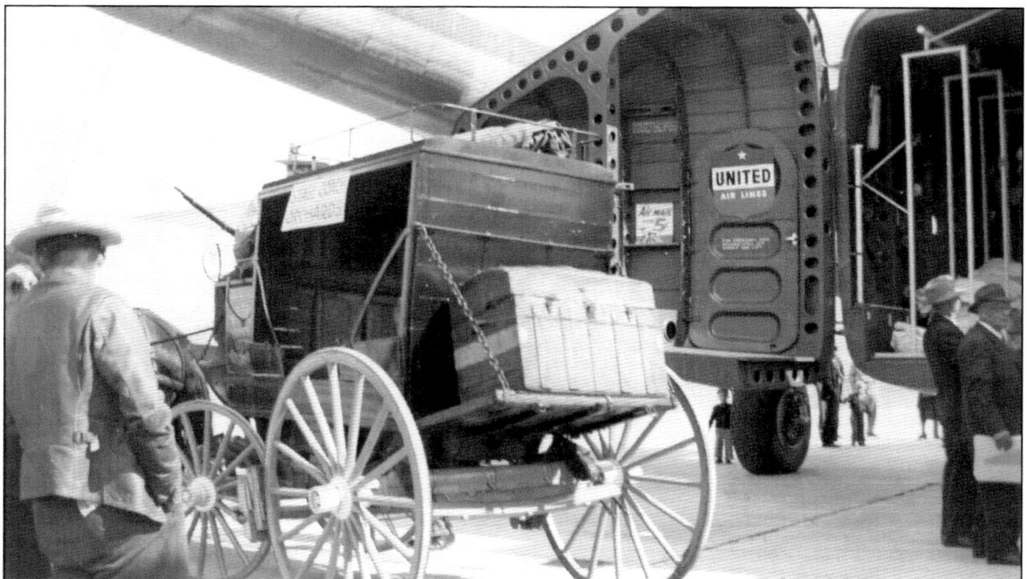

Old meets new at the Medford Airport. When the United Airlines Flying Mail Car arrived at the Medford airport, on hand to greet it was the Stage Coach Orchards' stage for a ceremonial transfer of mail. (SOHS No. 16790.)

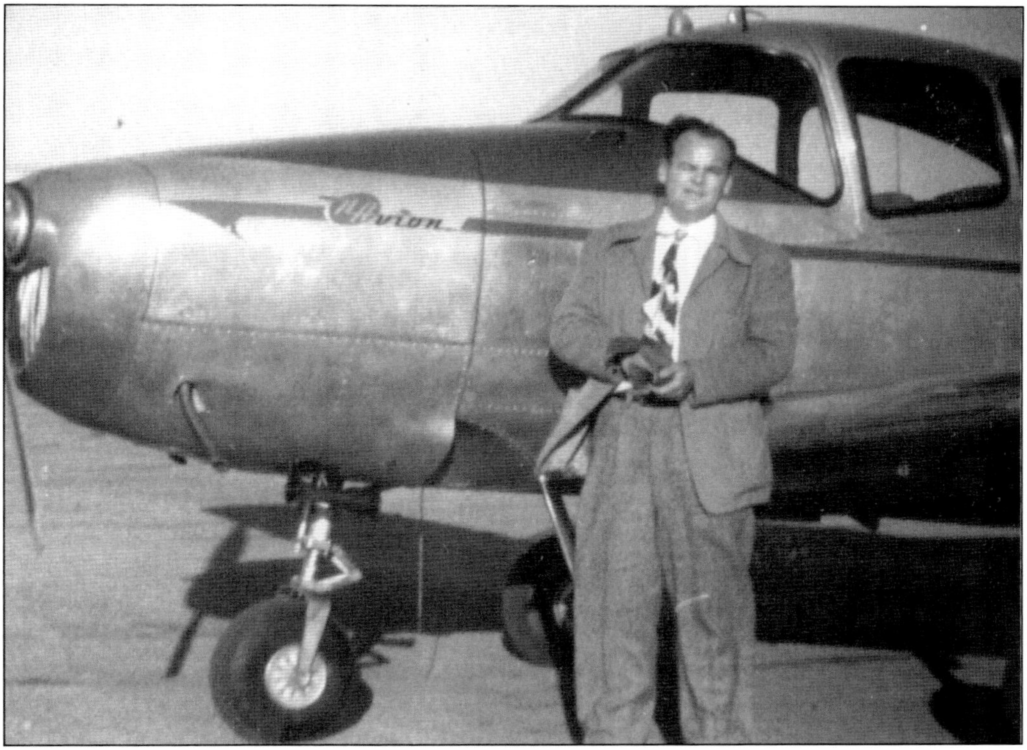

According to the *Medford Mail Tribune*, shortly after the conclusion of the war, Bob Alley "joined the ranks of the flying lumbermen" when he completed his flight training. Here he stands beside the four-place North American Navion that he and his brother Dick purchased in 1946. Alley is filling his ubiquitous pipe from a seemingly ever-present tobacco pouch. (Author's collection.)

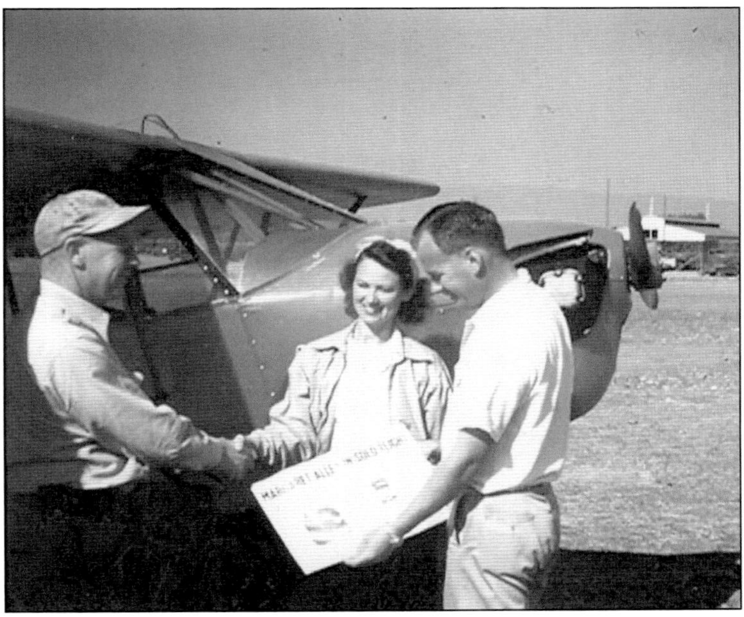

During the Second World War, what little civil aviation was permitted was transferred to the old Newell Barber Field site by the Jackson County Fairgrounds. Air space for nonmilitary flights was restricted to the confines of the Rouge Valley. Margaret Alley, wife of Dick Alley, earned her pilot's license sometime during this period or shortly after. One of the fair buildings is visible in the background. (Author's collection.)

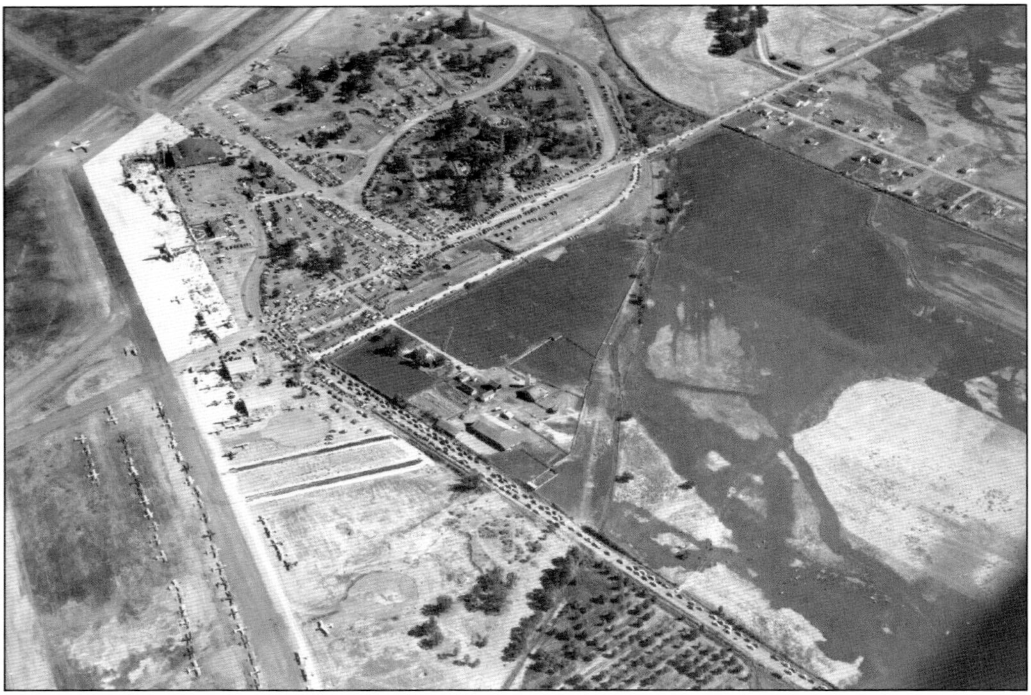

The public descended en masse on the municipal airport in September 1949. The occasion was the Aviation Fair commemorating the 20th anniversary of the opening of the new field in 1929. It was also the kick off for a new bond appeal to further improve airport facilities. (Rogue Valley International-Medford Airport.)

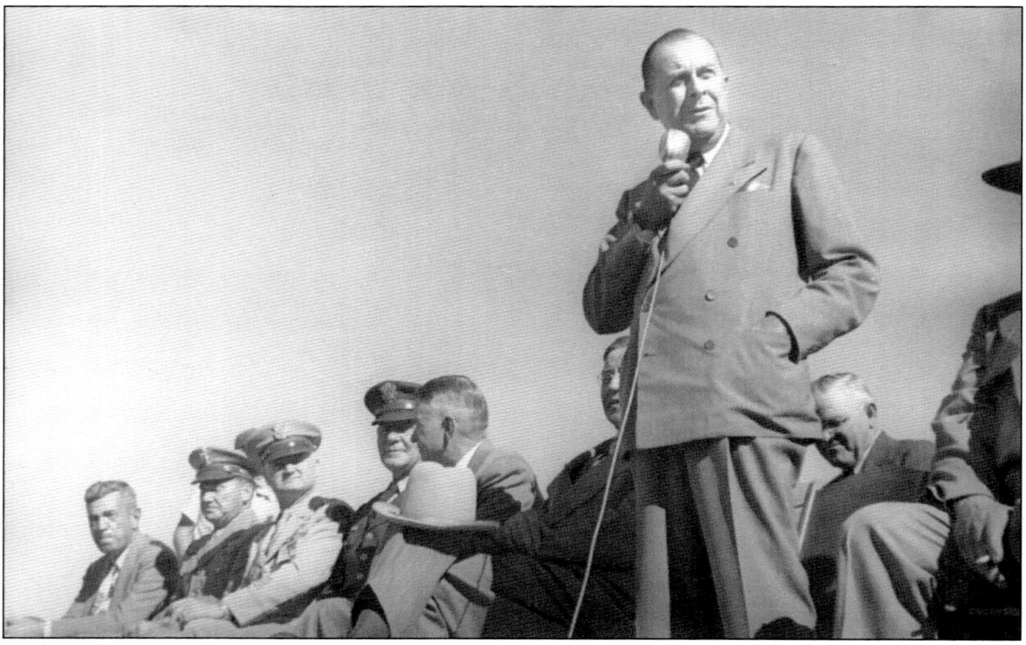

Seely Hall was Medford's pioneer aviator. In September 1949, he addressed the crowd at the Aviation Fair, celebrating the 20th anniversary of the Medford Municipal Airport. (SOHS No. 20708.)

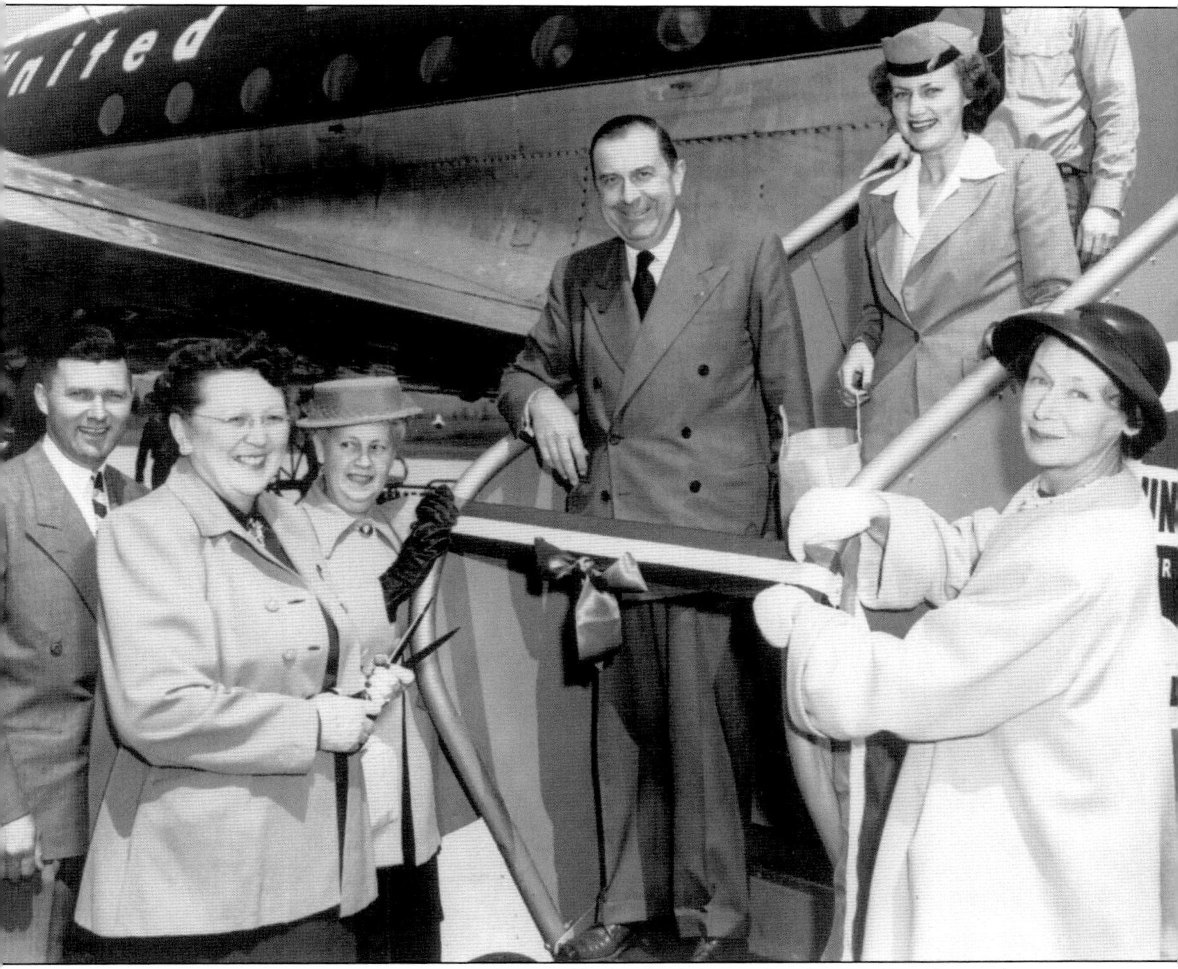

United Airlines introduced the DC-4 to the Medford schedule on April 30, 1950. Shown here is Fay Flynn, wife of Medford mayor Diamond Flynn, preparing to cut the ribbon welcoming Seely Hall, the first passenger, off the new plane. Hall was a pioneer in commercial aviation in the Rogue Valley and was United Airlines' vice president of ground operations. (SOHS No. 15321.)

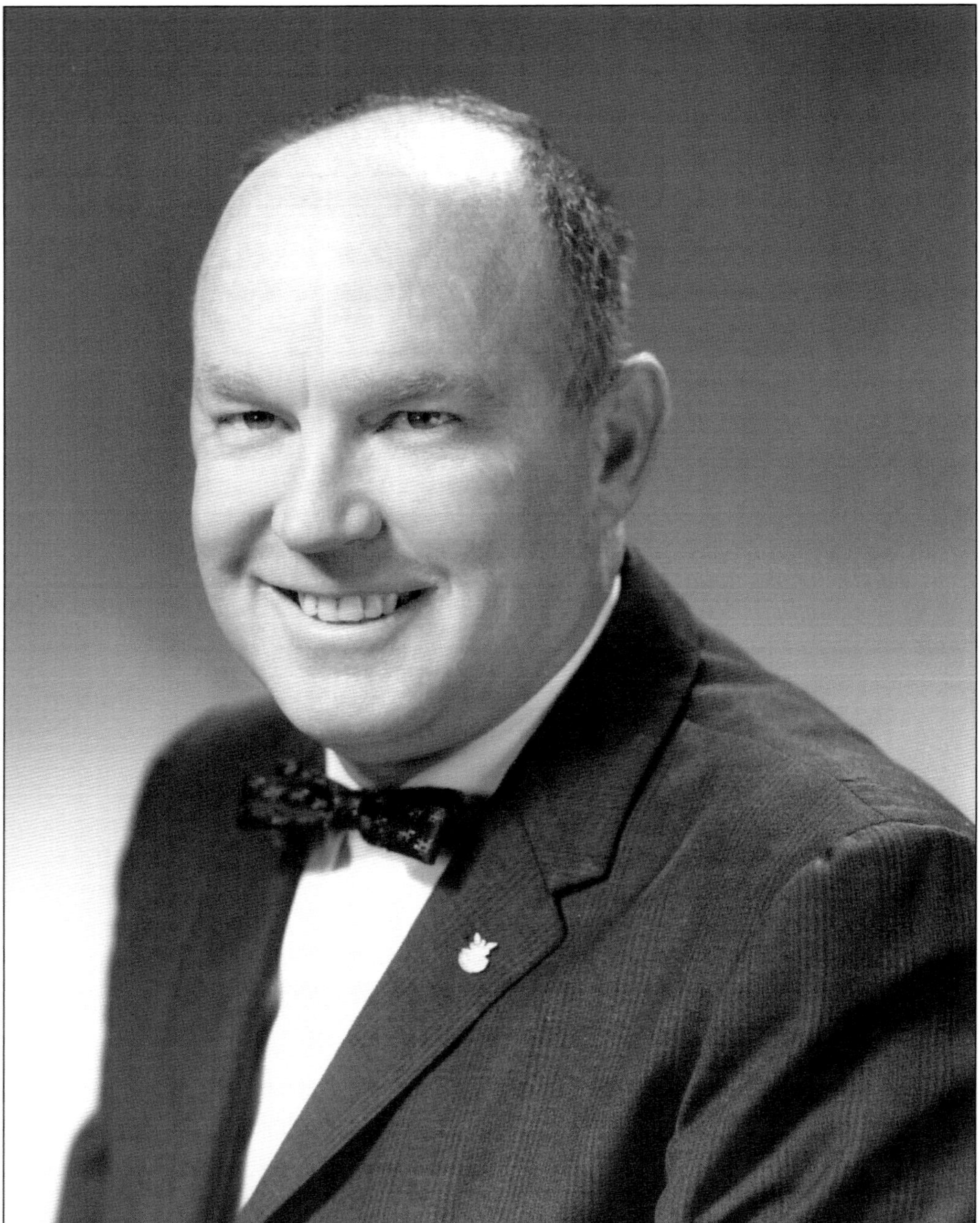

In 1949, George Milligan was an air traffic controller at the Medford airport. That year, the scourge of polio struck his good friend Art Winetrout, who was unable to survive the long drive to Portland for treatment. This loss galvanized Milligan's resolve to establish Mercy Flights, the nation's first nonprofit air ambulance run entirely by volunteers. (SOHS No. 16477.)

Patient Fred Hale is loaded into one of Mercy Flights' Twin Beech aircraft for transport in 1955. Founder George Milligan faces the camera. (SOHS No. 4511.)

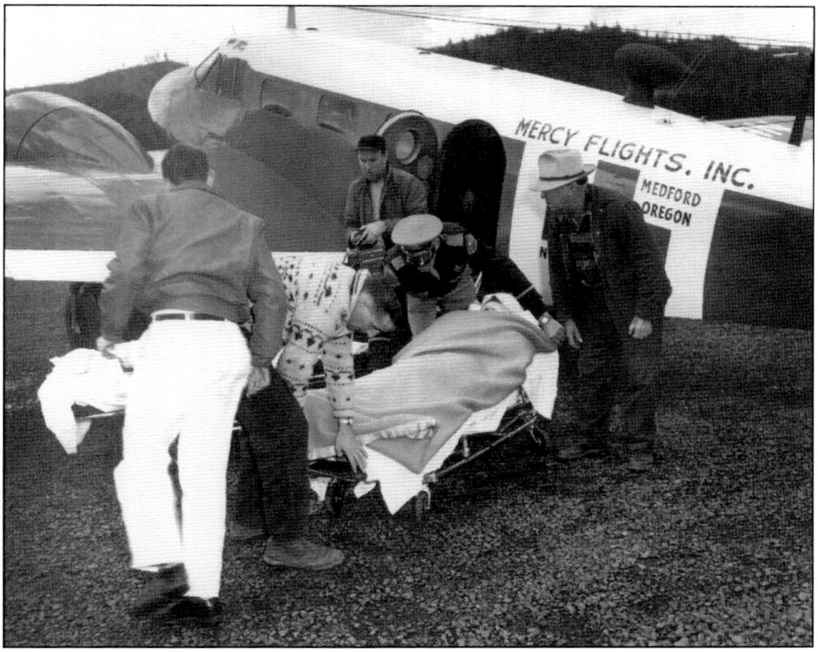

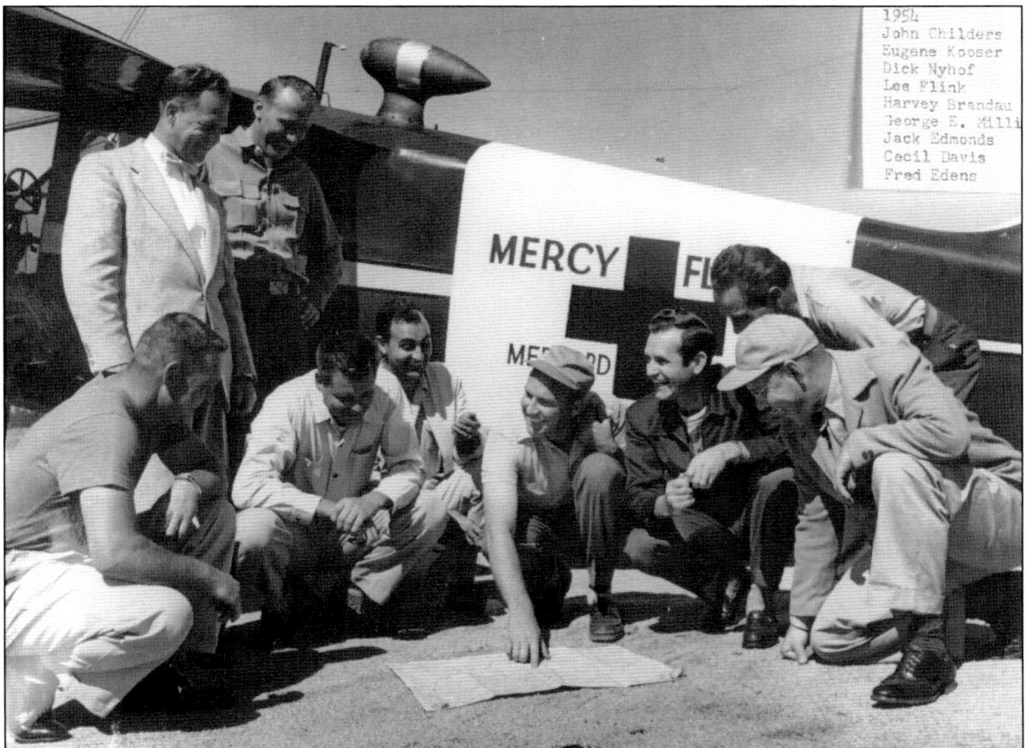

Mercy Flights founder George Milligan, center, shares a light moment with his volunteer pilots in 1954. Also photographed are, from left to right, John Childers, Eugene Koser, Dick Nyhoff, Lee Flink, Harvey Brandau, Jack Edmonds, Cecil Davis, and Fred Edens. (SOHS No. 16411.)

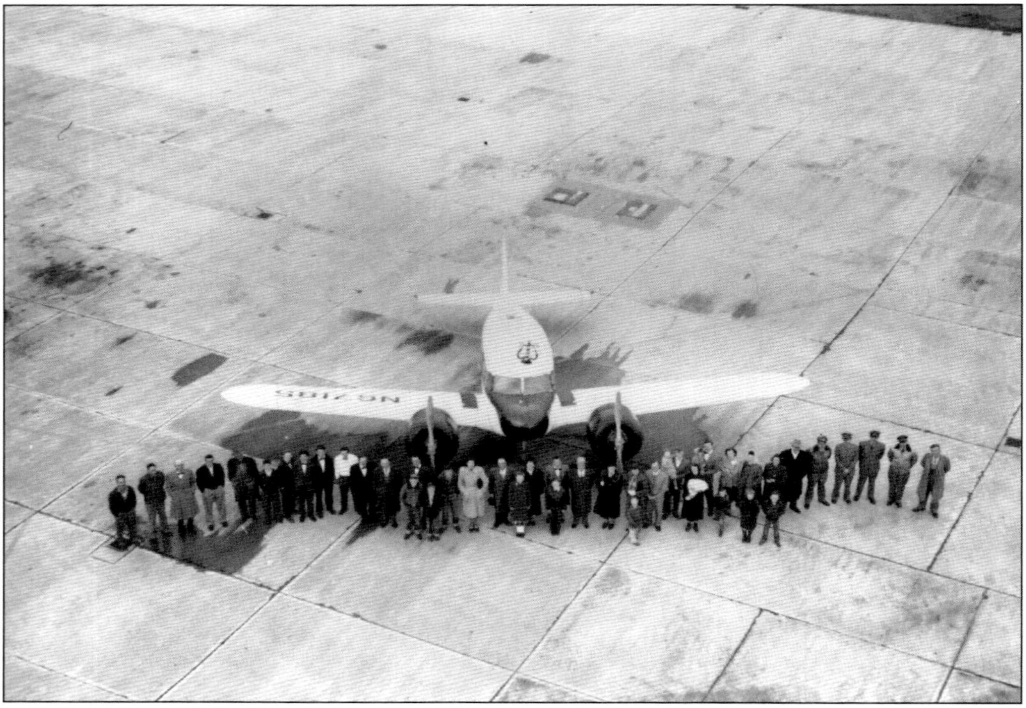

Mercy Flights' first aircraft was a surplus Cessna T-50 Bobcat, commonly referred to as the "Bamboo Bomber." This plane was christened *Rogue's Wings of Mercy* and transported its first patient on February 28, 1950. Fifty people, representing Mercy Flights' first 50 patients, pose in front of the plane at the Medford Municipal Airport. (SOHS No. 4741.)

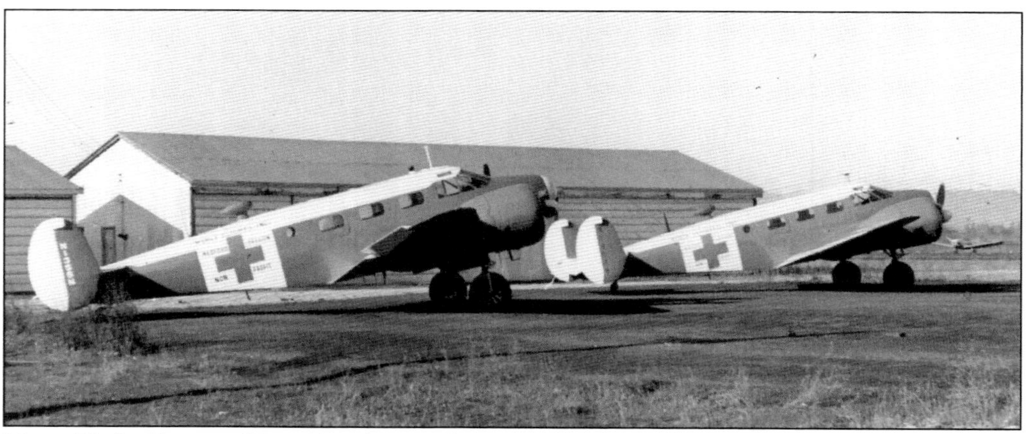

Mercy Flights' brace of Twin Beech aircraft sits on standby at the Medford Airport around 1955. Mercy Flights' planes were the only civilian aircraft of their day authorized to display the American Red Cross insignia. (SOHS No. 4737.)

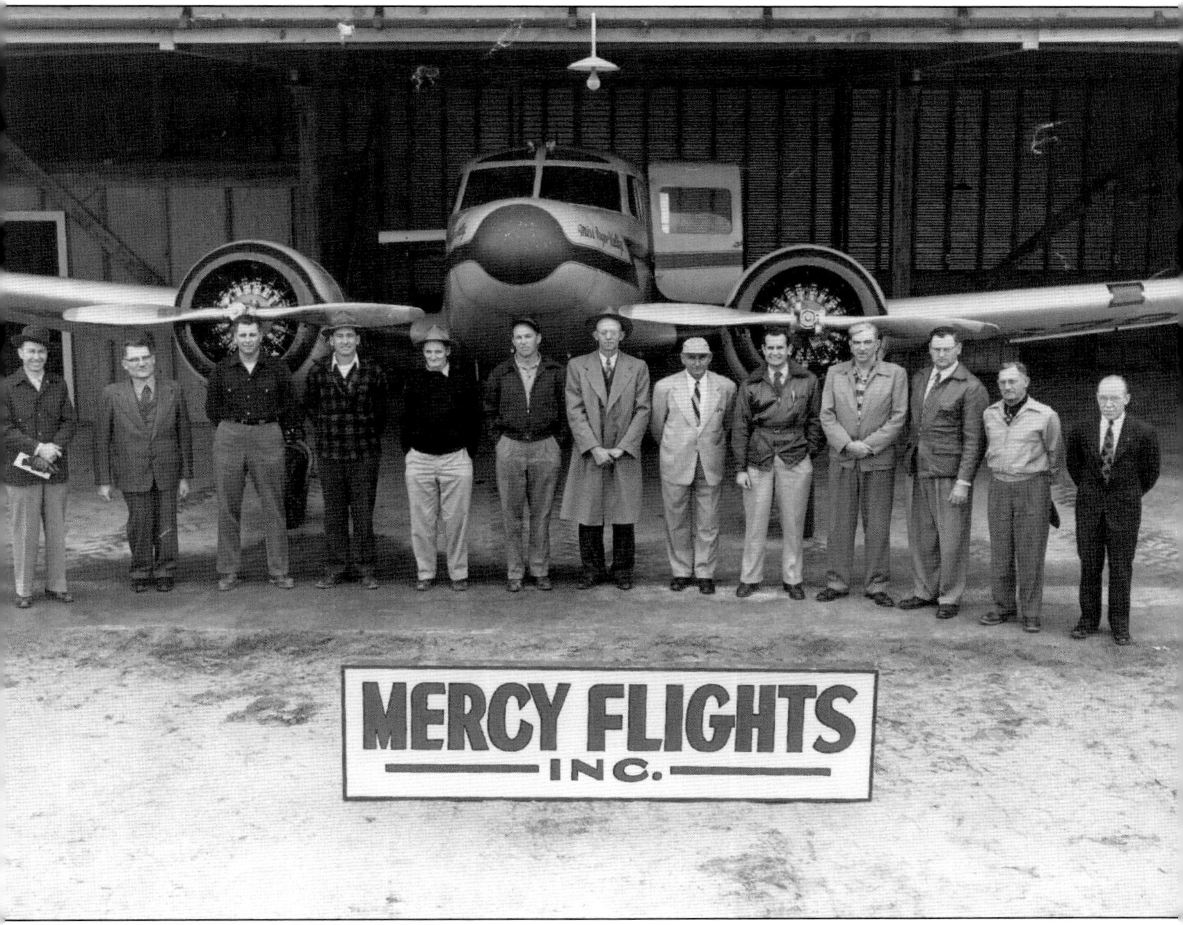

The volunteer pilots, mechanics, and support staff of Mercy Flights pose for a promotional photograph in front of their plane *Miss Rogue Valley*. (SOHS No. 20737.)

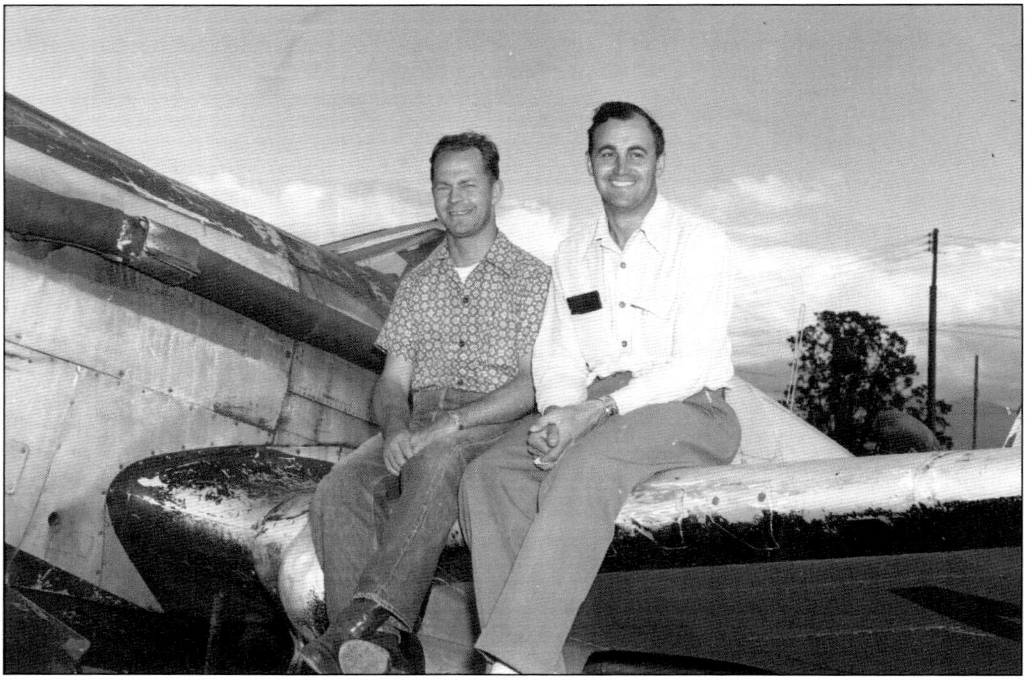

One of the primary disadvantages of aviation in southern Oregon was the thick valley fog that often plagued the area. In the years immediately after the war, George Milligan, the local air traffic controller and founder of Mercy Flights, along with others such as Gene Kooser (left) and Harvey Brandau (right), successfully experimented with cloud seeding—a form of weather modification—using dry ice to clear the fog. (SOHS No. 20715.)

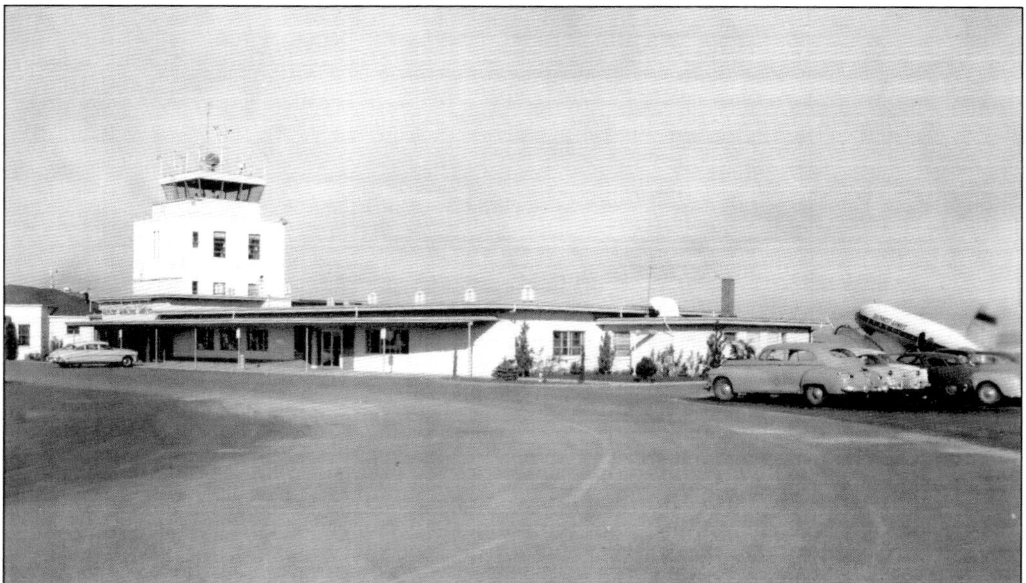

After 20 years, it was clear that the 1929 airport building had become inadequate and obsolete. By 1953, a modern new facility had been constructed, complete with an up-to-date control tower. (Rogue Valley International-Medford Airport.)

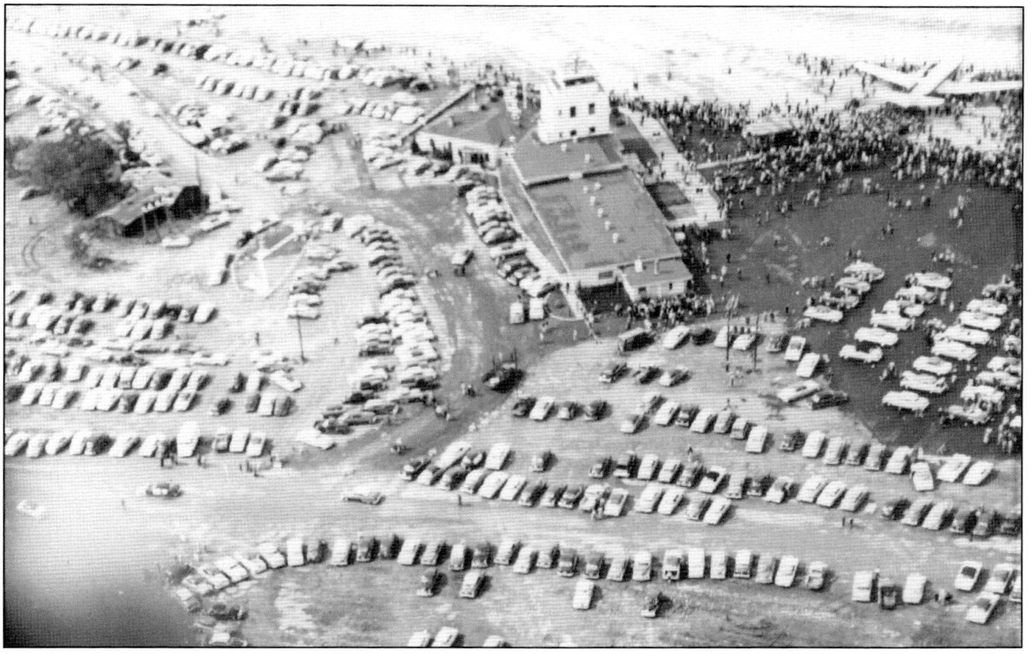

A large dedication ceremony was held for the June 7, 1953, opening of the newest airport terminal and administration building. The new terminal included a Civil Aviation Authority tower, a U.S. Weather Bureau facility, a waiting room, and ticketing for the three airlines that serviced Medford: United Airlines, Southwest Airlines, and West Coast Airlines. The portion of the building jutting out on the left is the United Airlines terminal built during the war. (Rogue Valley International-Medford Airport.)

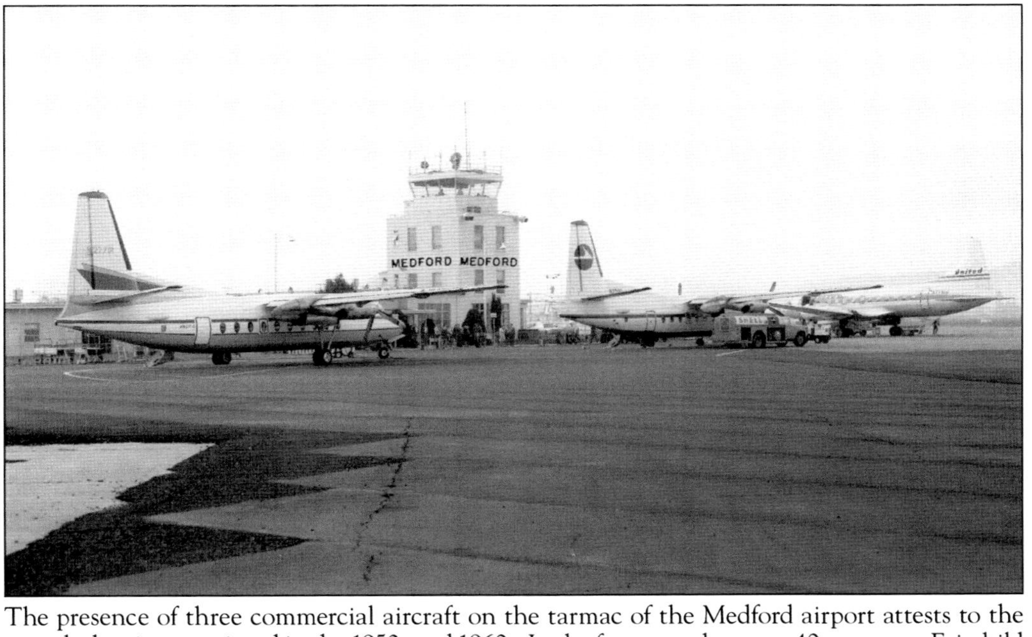

The presence of three commercial aircraft on the tarmac of the Medford airport attests to the growth the airport enjoyed in the 1950s and 1960s. In the foreground are two 40-passenger Fairchild F-27 planes, introduced in 1958. (Rogue Valley International-Medford Airport.)

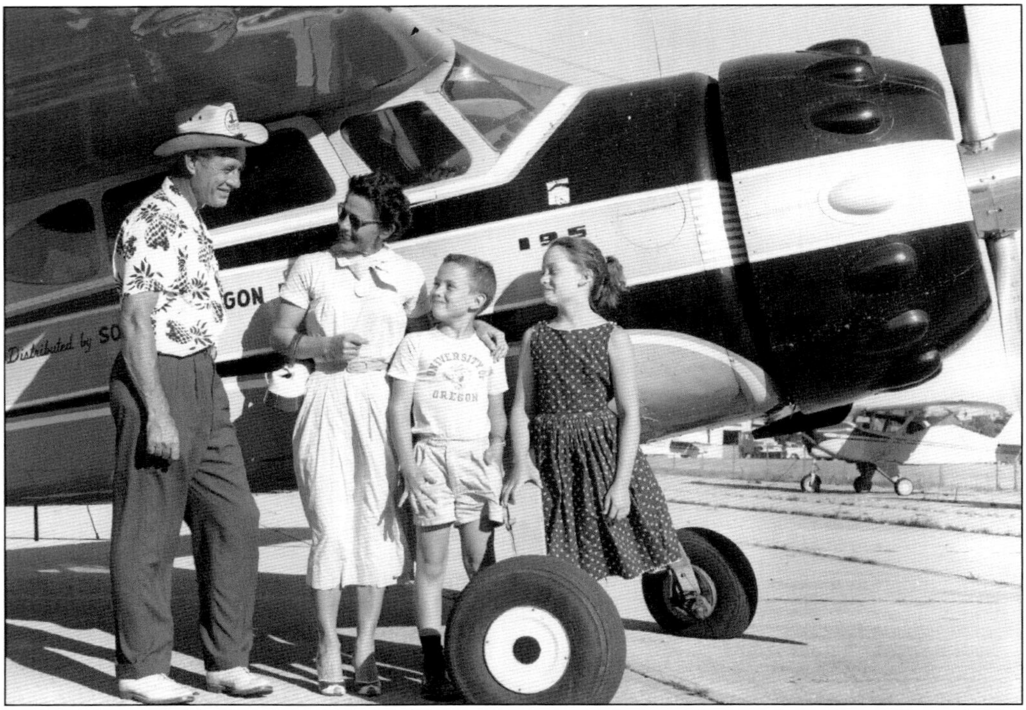

The postwar years enjoyed an expansion of business and family participation in general aviation. This family stands beside a Cessna 195 Business Liner, Cessna's last radial engine aircraft. Its roomy cabin provided comfort for a group of businessmen or families. It was in production between 1947 and 1954. (SOHS No. 14637.)

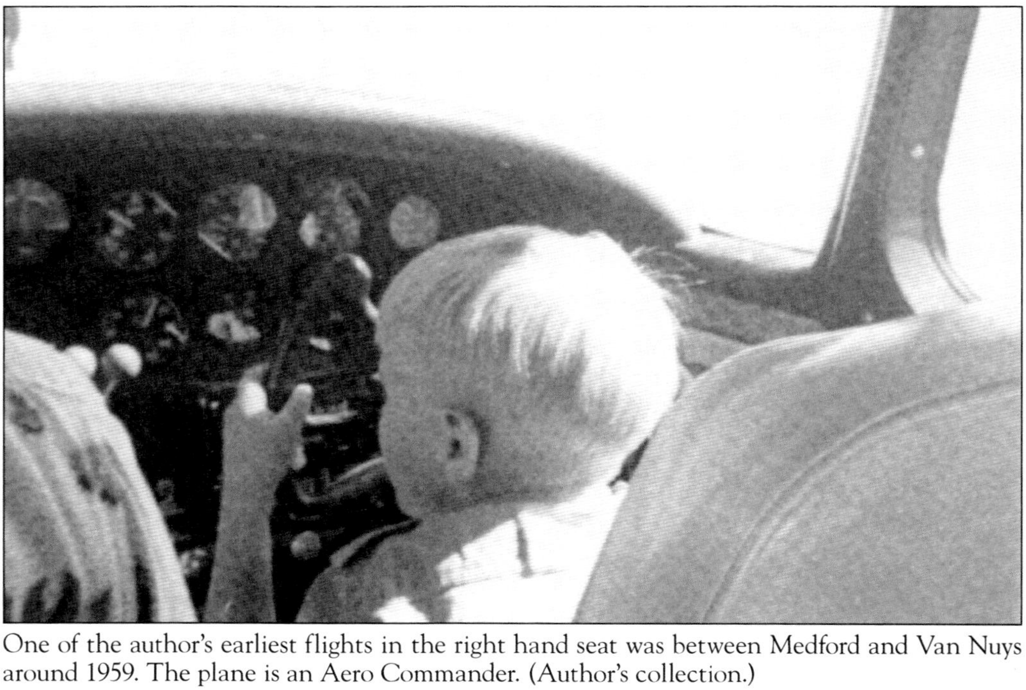

One of the author's earliest flights in the right hand seat was between Medford and Van Nuys around 1959. The plane is an Aero Commander. (Author's collection.)

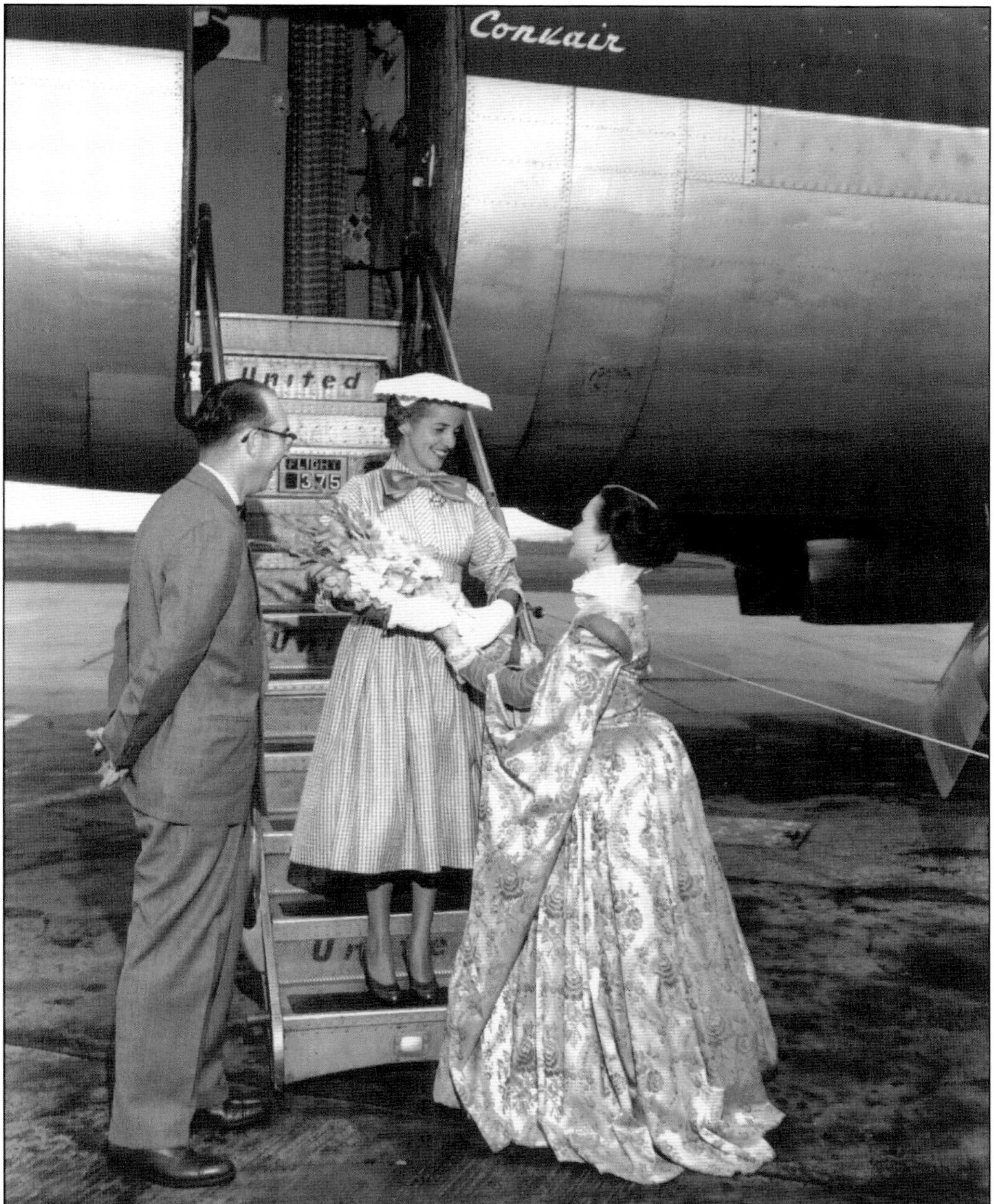

The popular radio program *Queen for a Day* became a hit television program between 1956 and 1964. Medford photographer Vern Shangle caught the arrival of this "Queen for a Day" upon her arrival in Medford. The woman welcoming her is probably an actress from the Ashland Shakespeare Company. (SOHS No. 19941.)

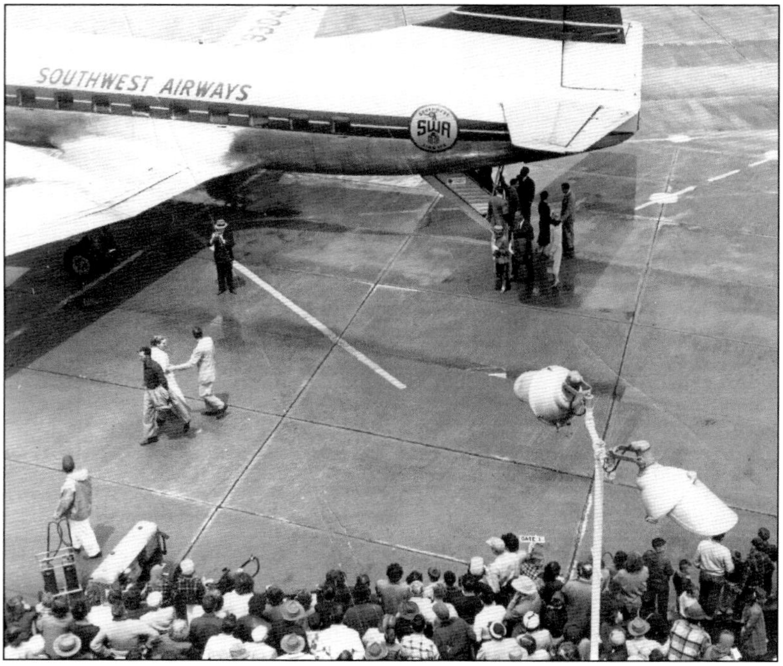

Southwest Airlines added Medford to its California routes in the early 1950s. This Martin 2-0-2 generated a lot of public interest around 1952. Perhaps the crowd gathered to witness its inaugural landing. (Rogue Valley International-Medford Airport.)

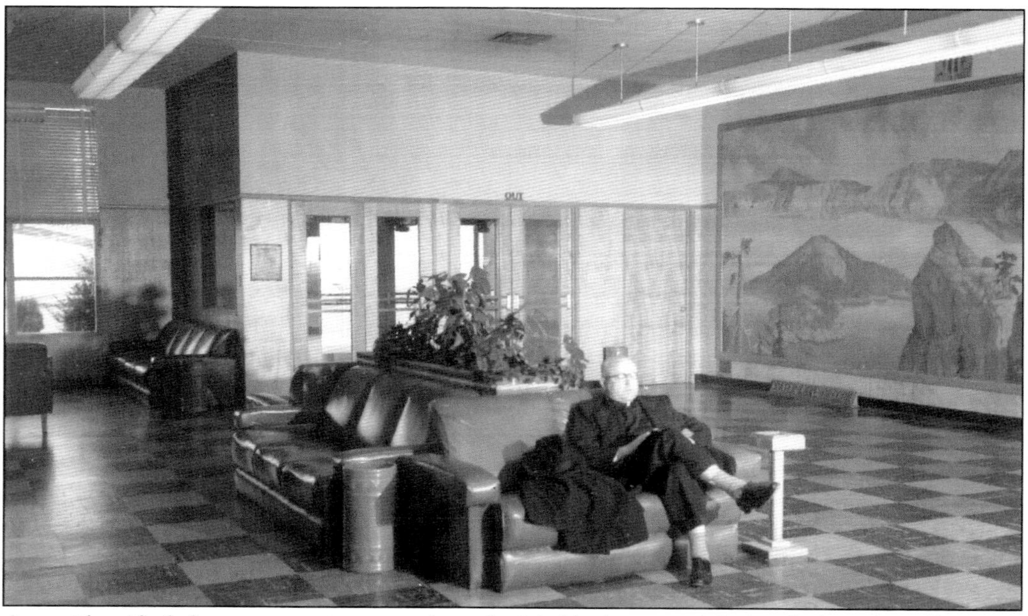

An unidentified clergyman waits patiently in the Medford terminal for his flight to be called. The plush leather sofas and chairs, an ashtray nearby, and the solitude of the terminal are reminders of an era of airplane travel now almost forgotten. (Rogue Valley International-Medford Airport.)

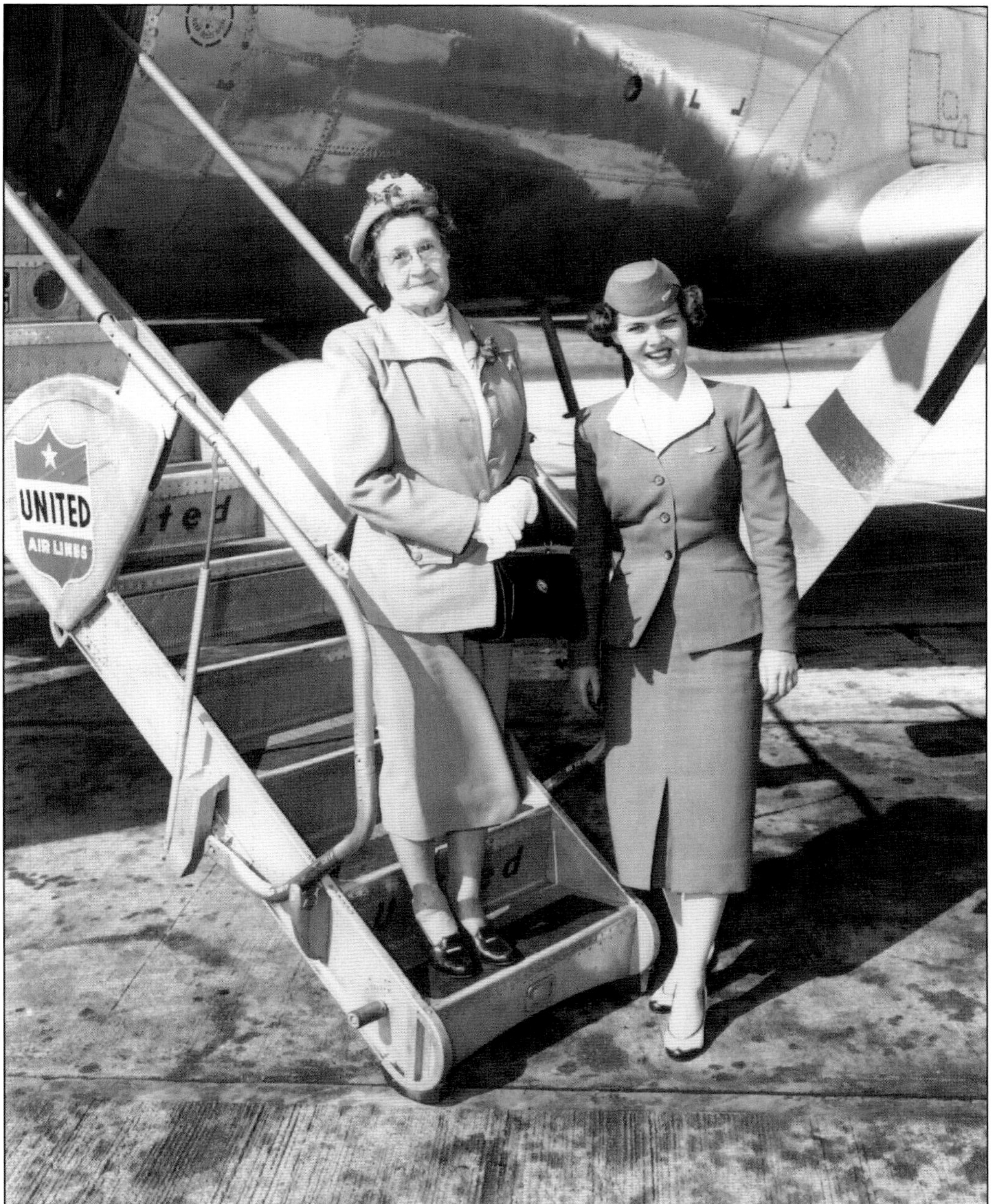

An unidentified 75-year-old woman prepares to board a United Airlines plane for her very first airplane flight around 1958. The local newspaper considered this event significant enough to dispatch a photographer to record it. (SOHS No. 19739.)

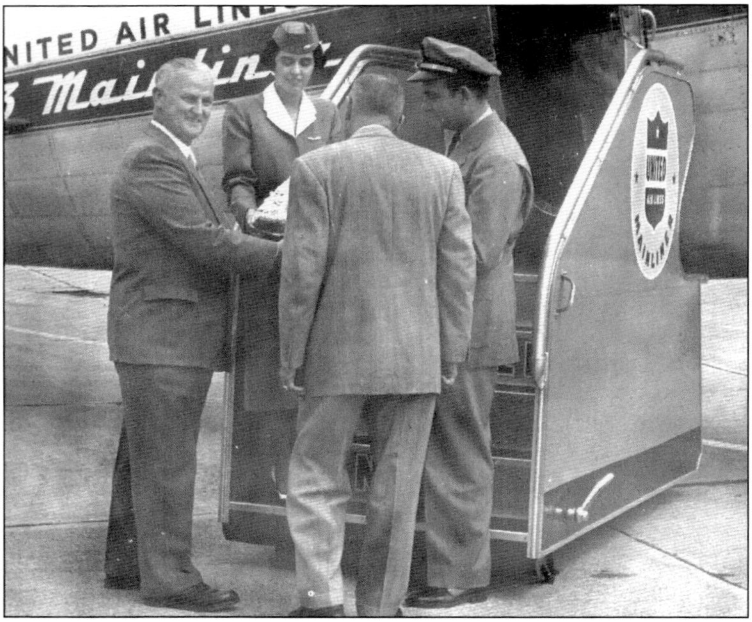

If the photographic record is any indication, it seemed fairly common for distinguished visitors to be welcomed at the airport with a gift basket from Harry and David, the Rogue Valley's iconic gift fruit vendors. (Rogue Valley International-Medford Airport.)

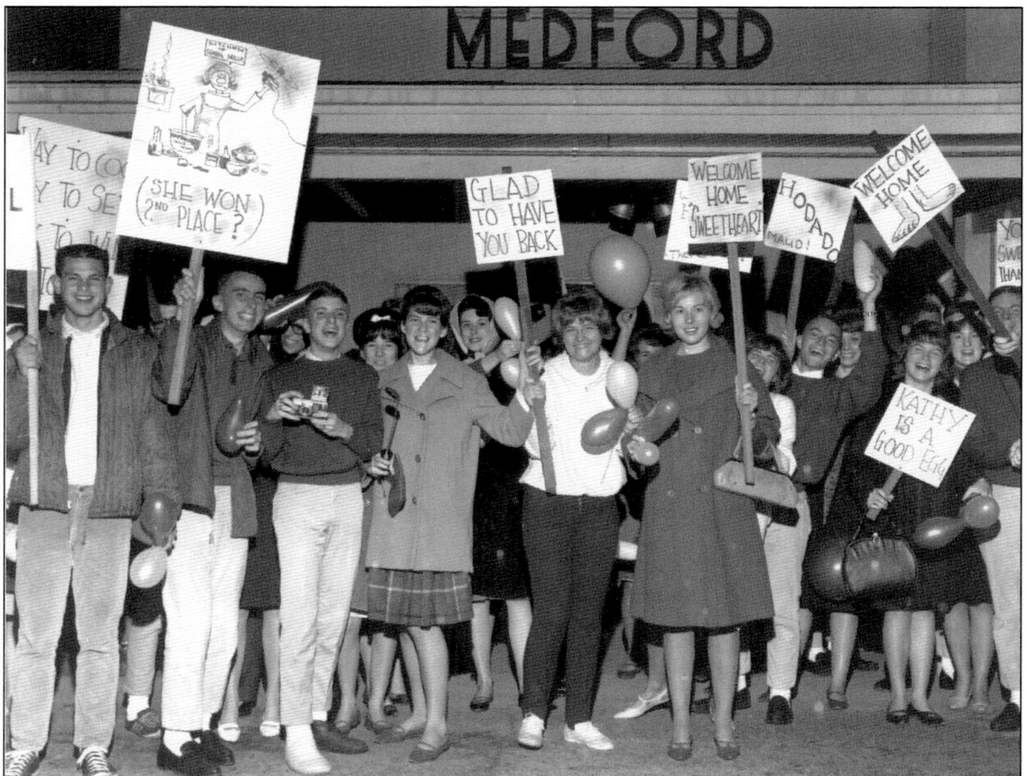

Kathryn Ann Foley won second place in the 1964 Betty Crocker All-American Homemaker of Tomorrow competition in New York City. She was welcomed home by a gathering of fellow Medford High School students at the Medford Airport. (SOHS No. 15344.)

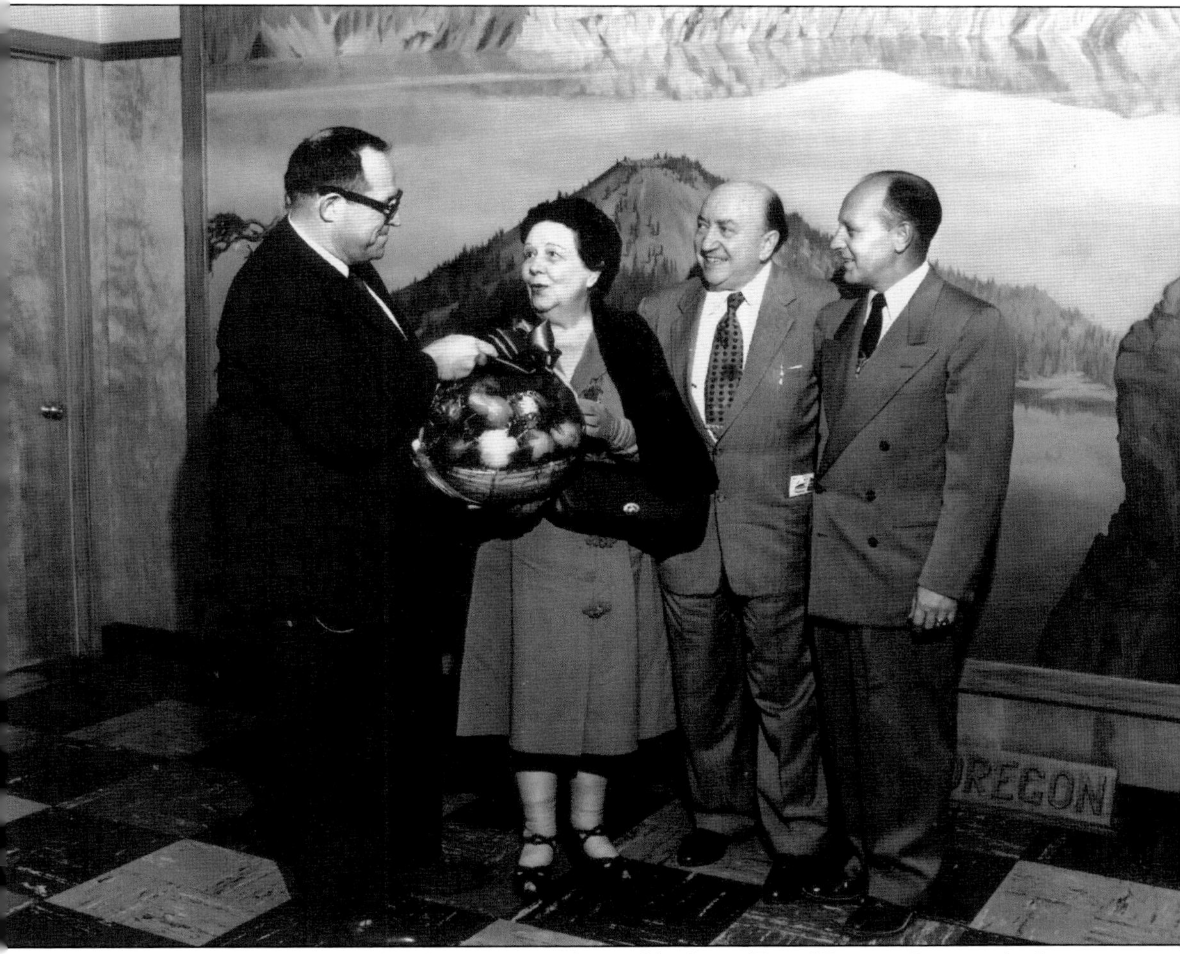

This woman is being presented with a Harry and David basket of Royal Riviera Pears and other assorted fruits in the lobby of the Medford terminal. Behind her is the large painting of Crater Lake that had long welcomed visitors to the airport. The painting is now preserved in the collections of the Southern Oregon Historical Society. (SOHS No. 19942.)

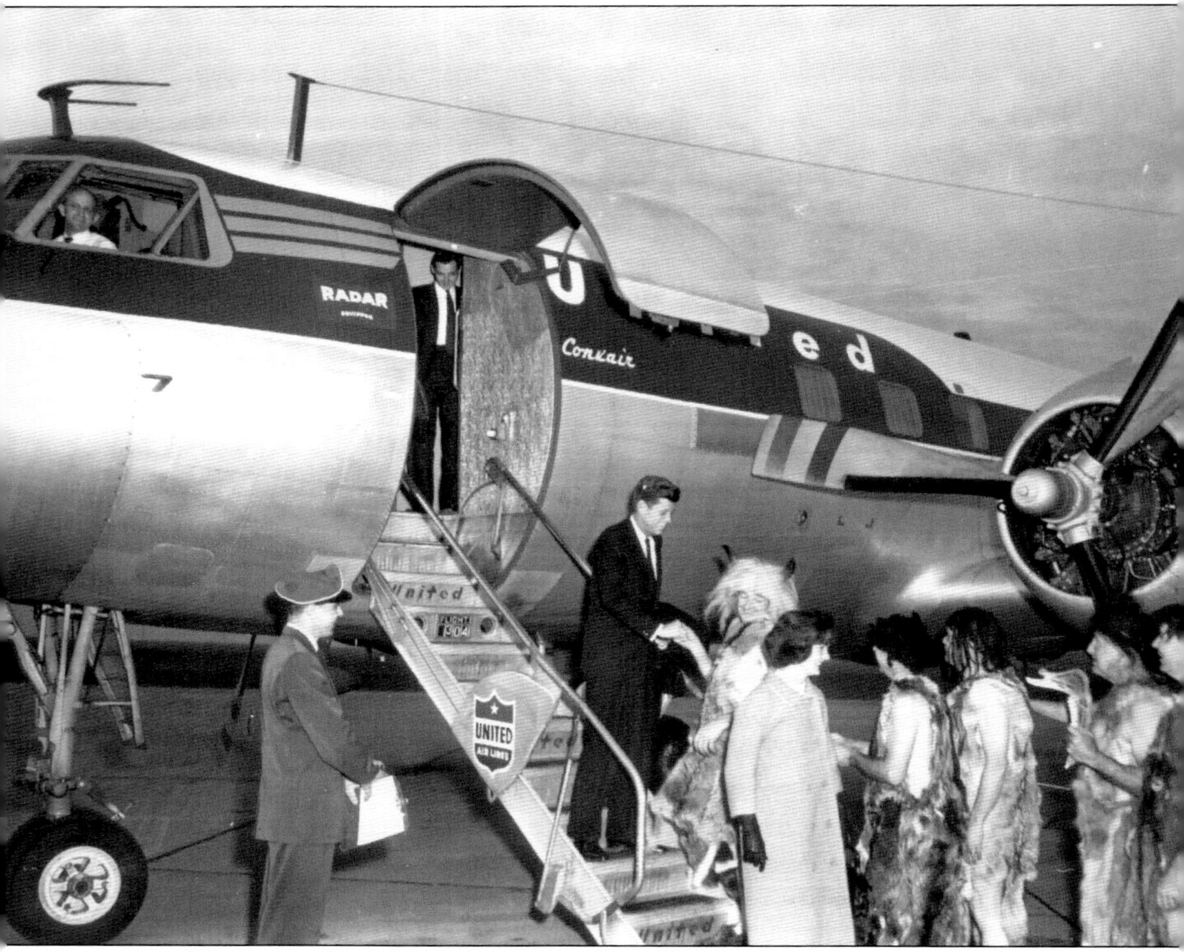

Sen. John F. Kennedy visited Medford in 1959 while laying the groundwork for his presidential bid the following year. He is seen here following his wife off the plane and being welcomed by members of the Grants Pass Cavemen, a civic organization famous for their animal skins and clubs. (SOHS No. 14066.)

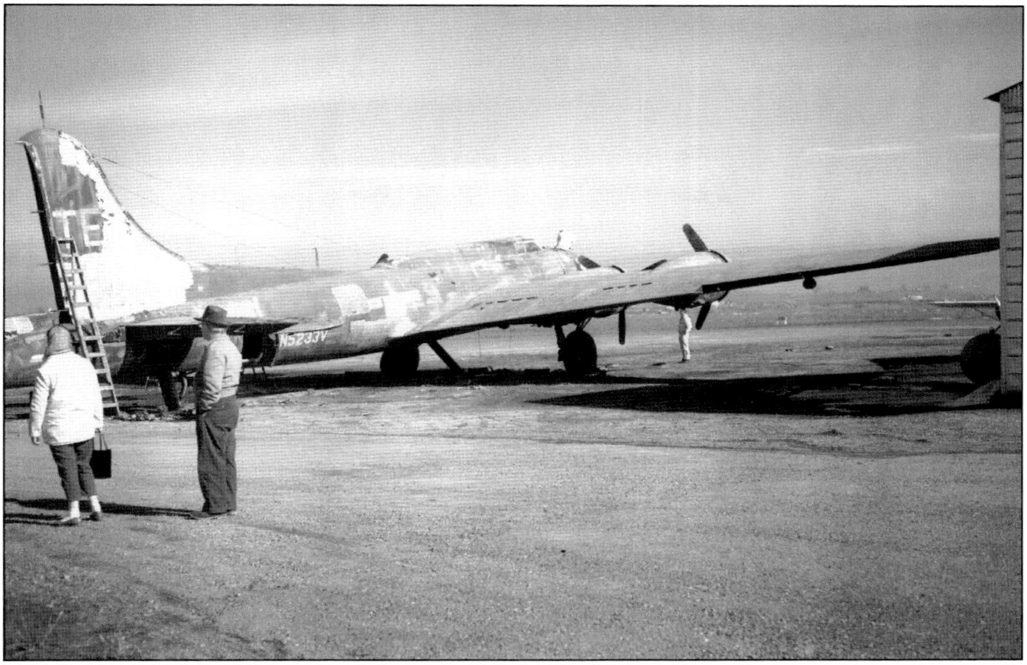

This B-17 was rebuilt and converted into a tanker at the Medford airport in the early 1960s and operated out of Medford between 1962 and 1969. It was later restored and placed on exhibit in the Air Force Armaments Museum at Elgin AFB in Florida. (Photograph ©Terry Skibby.)

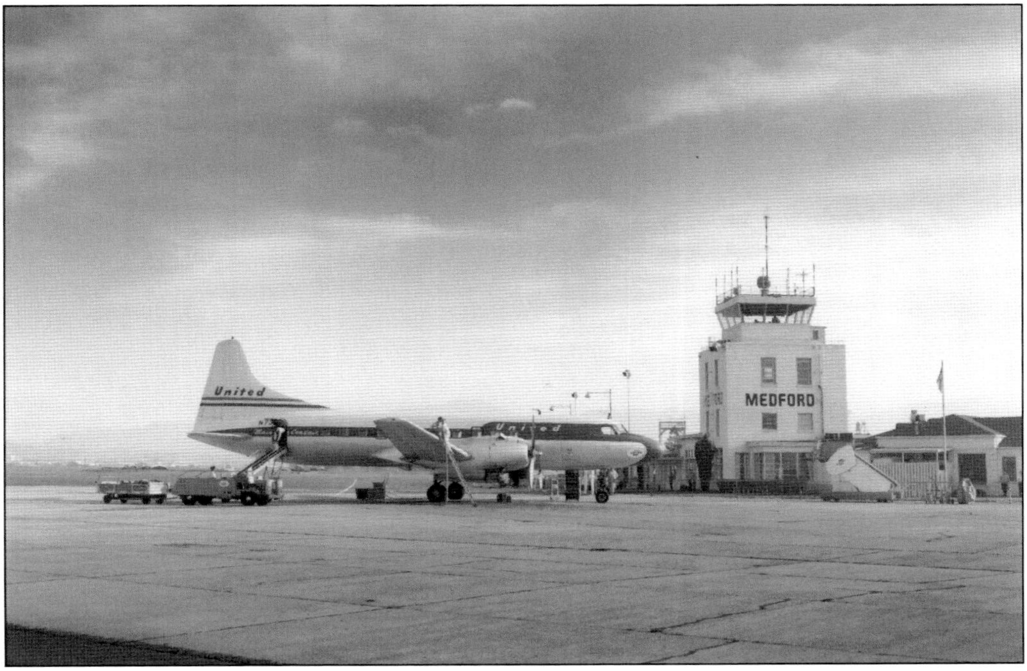

United Airlines introduced the Convair to the Medford run around 1952 and operated them until 1968. In this photograph, ground crew members prepare the last Convair out of Medford in February 1968. (SOHS No. 20734.)

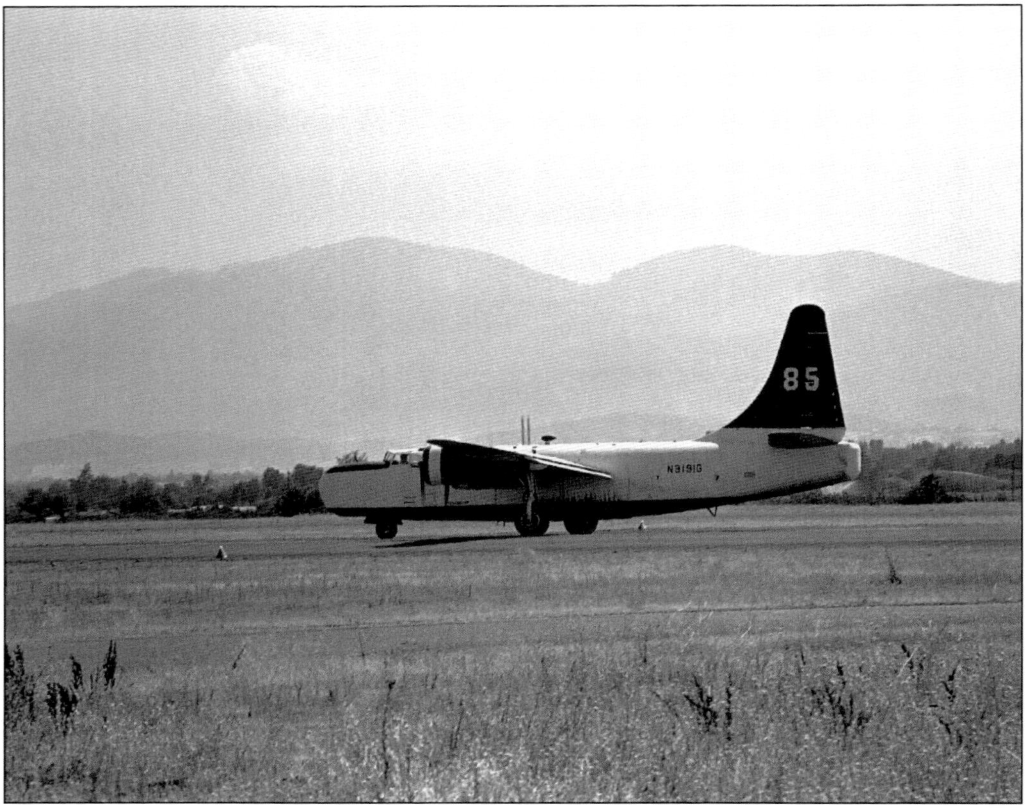

Medford's first airport was built to accommodate forest fire patrols. That role continues with the presence of the tanker base at Medford for firefighting. This PB4Y Privateer tanker operated out of Medford until it was ditched and sank in Diamond Lake in 1970. (Photograph ©Terry Skibby.)

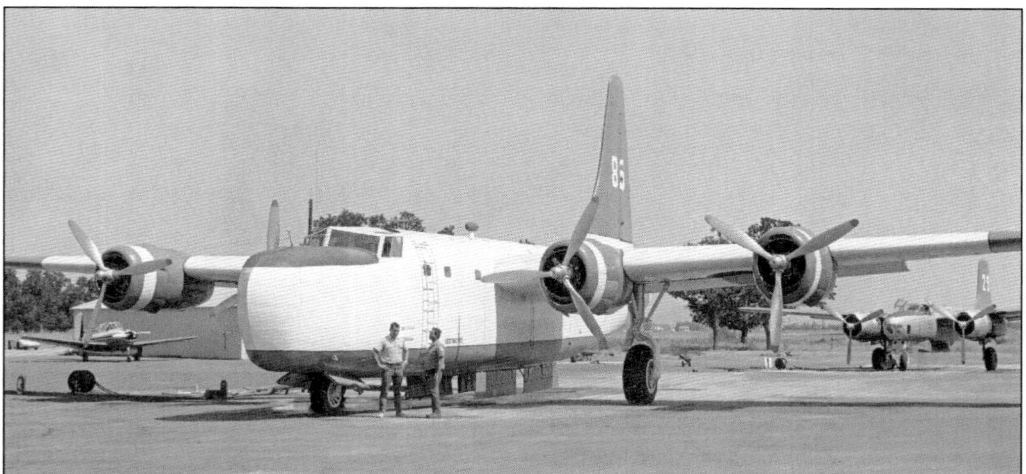

Pictured are two firefighting aircraft loading retardant at the tanker facility at the Medford airport around 1970. (Photograph ©Terry Skibby.)

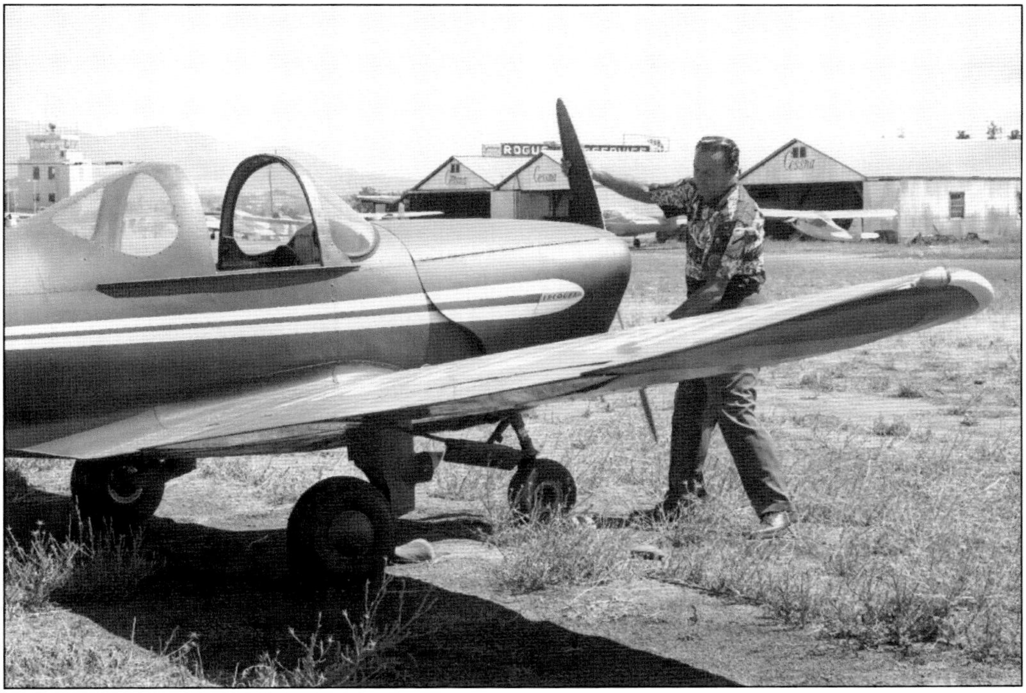

Civil aviation has long played a significant role in Medford aviation. This man is moving his Ercoupe, a popular small plane. In the background are a few of the many hangars devoted to general aviation; these belong to Rogue Air Service, an authorized Cessna dealer and service center. (SOHS No. 20705.)

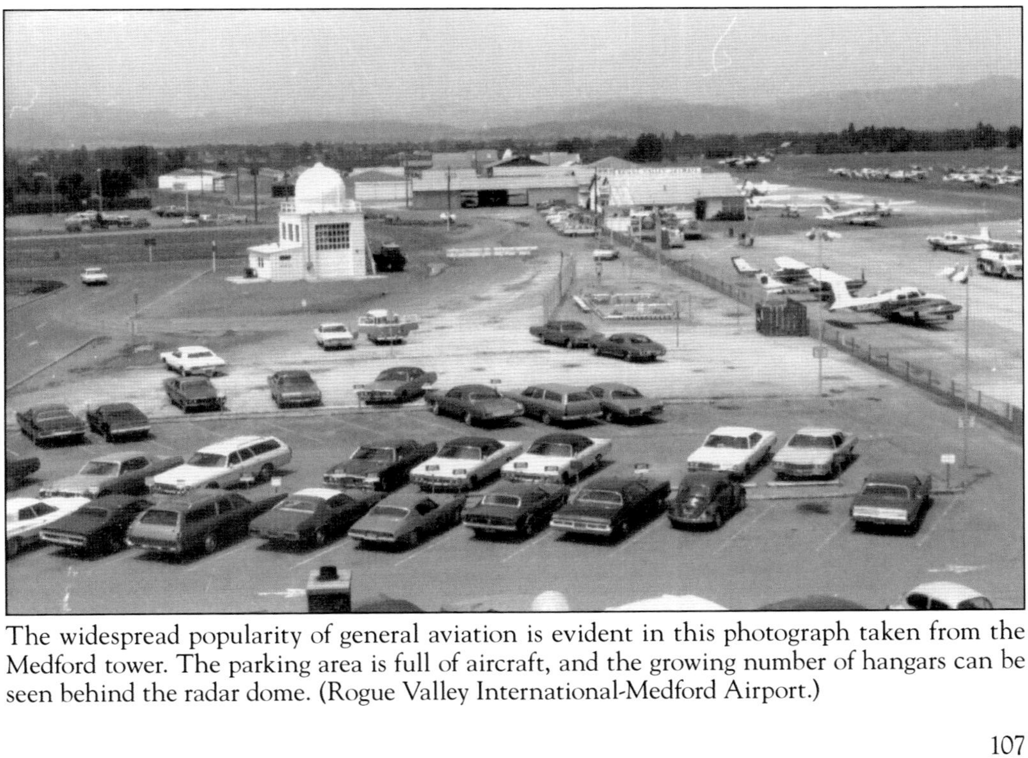

The widespread popularity of general aviation is evident in this photograph taken from the Medford tower. The parking area is full of aircraft, and the growing number of hangars can be seen behind the radar dome. (Rogue Valley International-Medford Airport.)

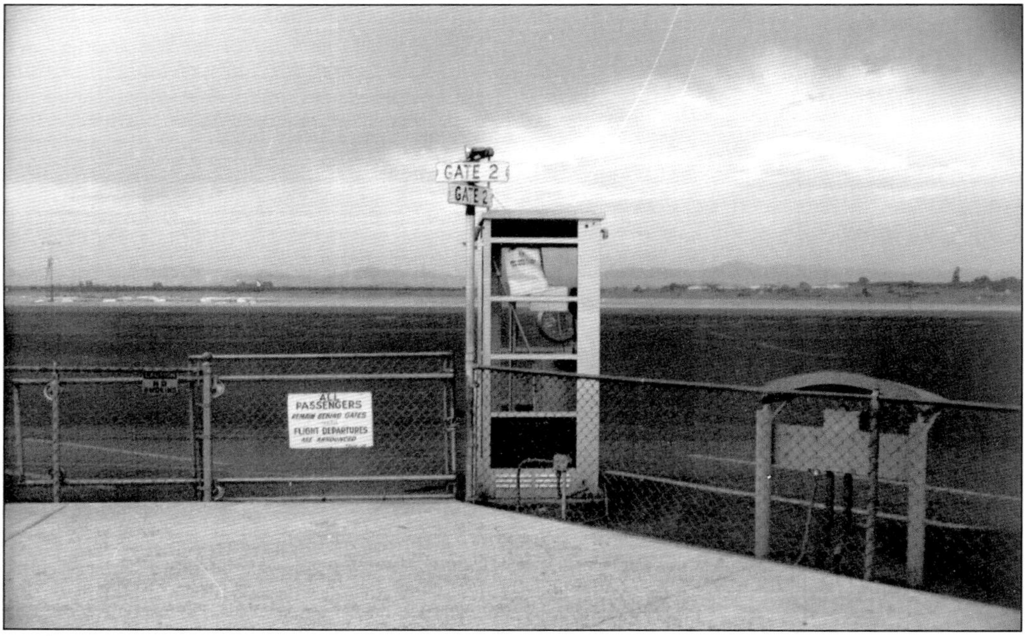

In the present age of jetways and cellular phones, an actual chain link fence and a telephone booth seem out of place at an airport. For most flights in and out of Medford, however, one can still enjoy the walk across the tarmac and a view of their aircraft before they board. As for the telephone booth, this one more likely contained emergency equipment. (Rogue Valley International-Medford Airport.)

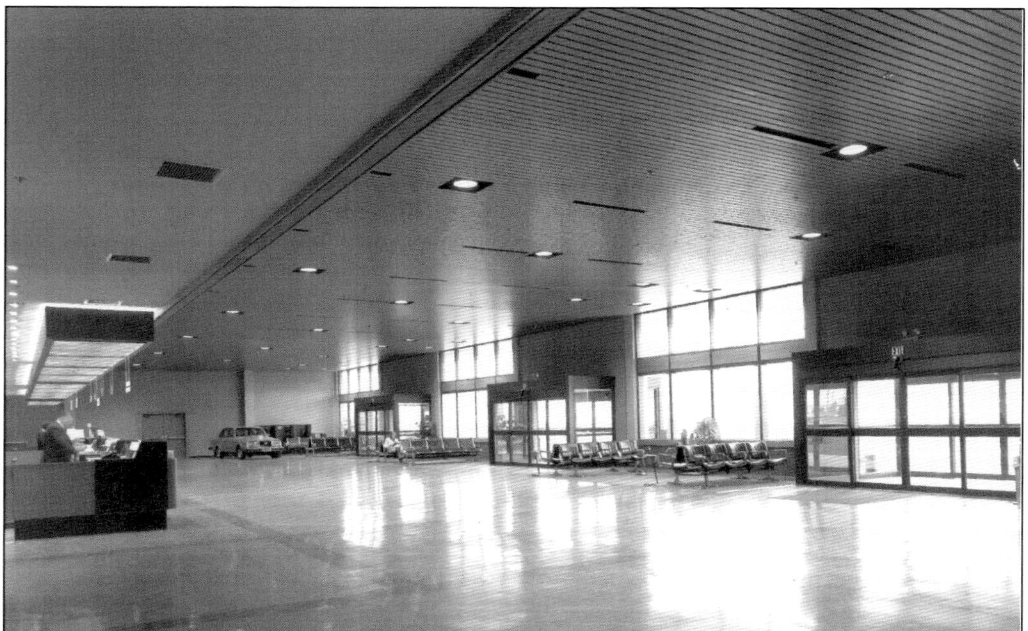

Here is the main lobby and ticketing area of the Medford terminal, as it appeared in the years before being razed in 2009. (SOHS No. 4962.)

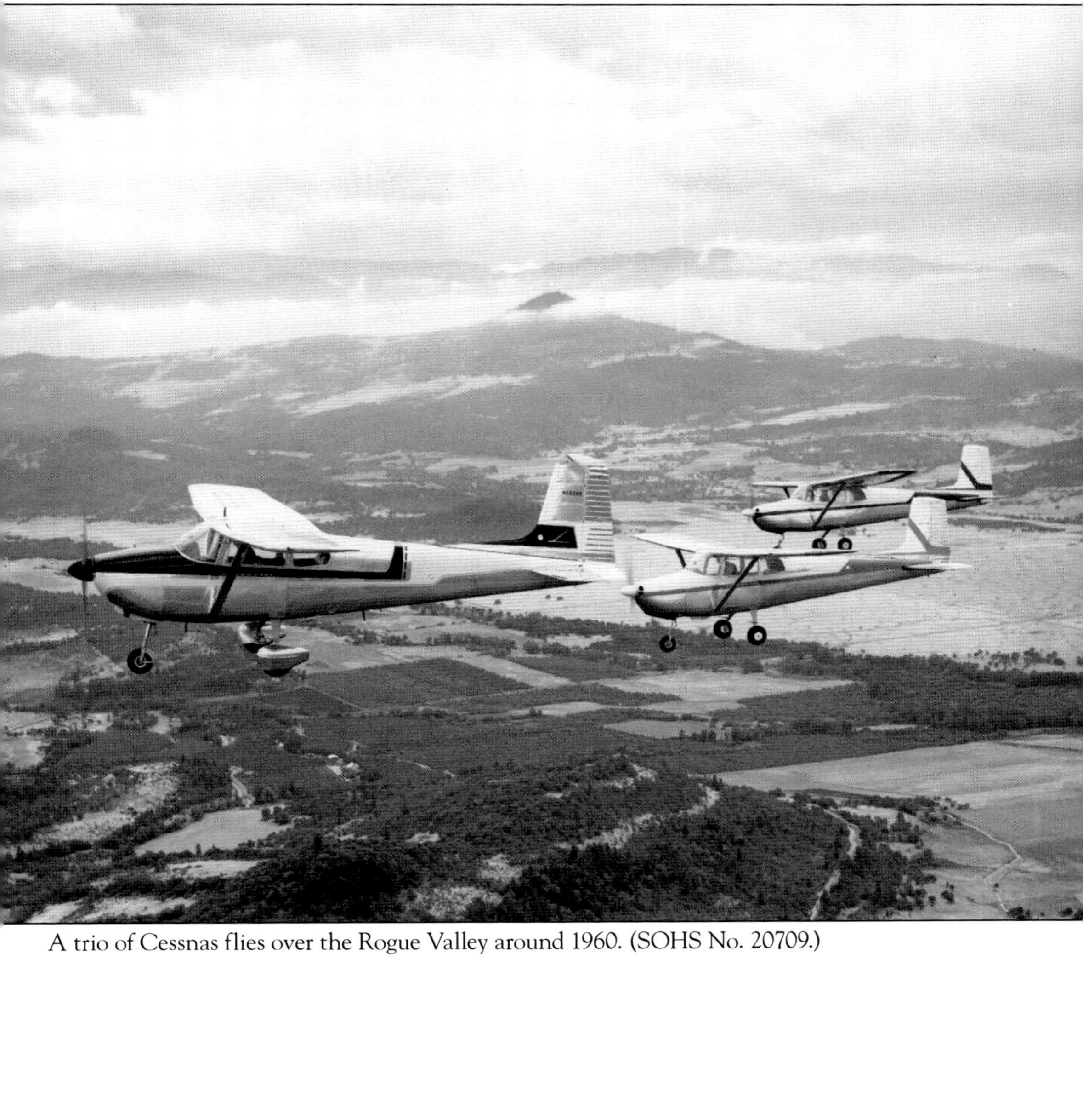

A trio of Cessnas flies over the Rogue Valley around 1960. (SOHS No. 20709.)

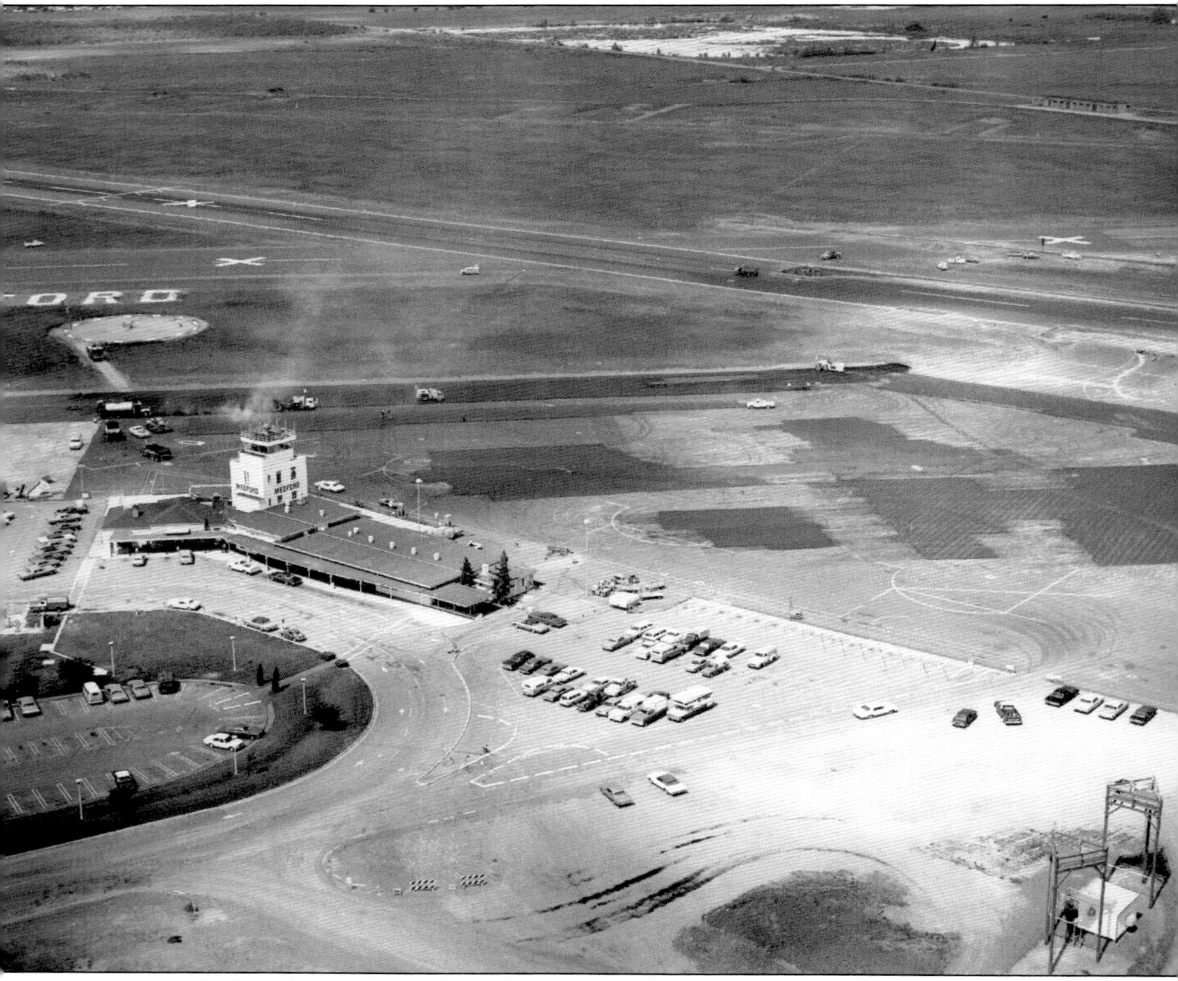

In 1971, the Medford airport undertook some major resurfacing work on the runways and taxiways. The large white Xs on the runways indicate that the field is closed to all traffic. (Rogue Valley International-Medford Airport.)

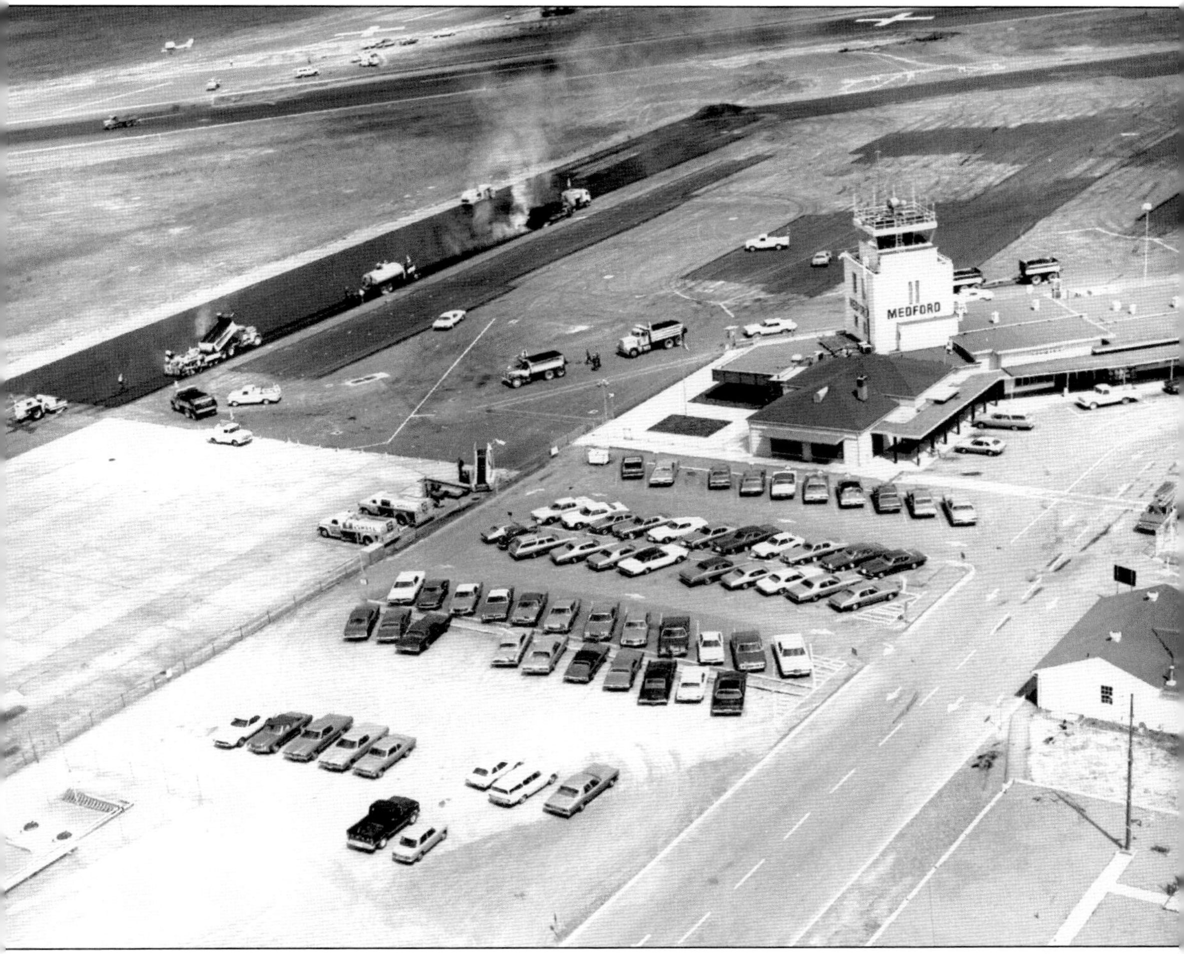

A major resurfacing project on the taxiway was made in 1971. This aerial of the work also shows clearly how the United Airlines terminal, constructed during the military lease of the airport, was incorporated into the new terminal building behind it. (Rogue Valley International-Medford Airport.)

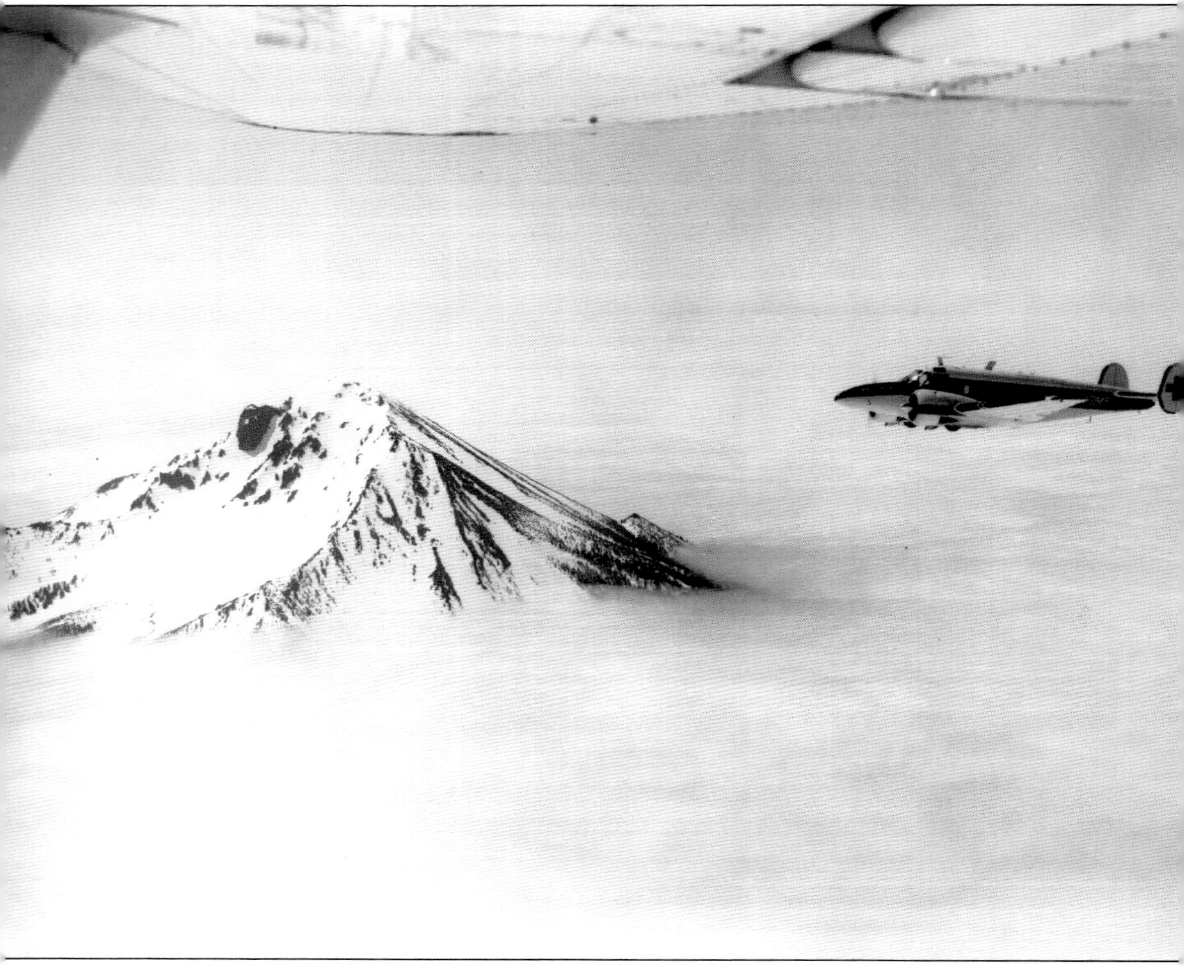

In 1983, Mercy Flights' newest aircraft, a Beechcraft H-18, flies past southern Oregon's volcanic landmark, Mount McLaughlin. The low clouds below give an indication of just how thick a Rogue Valley fog can be. (SOHS No. 16412.)

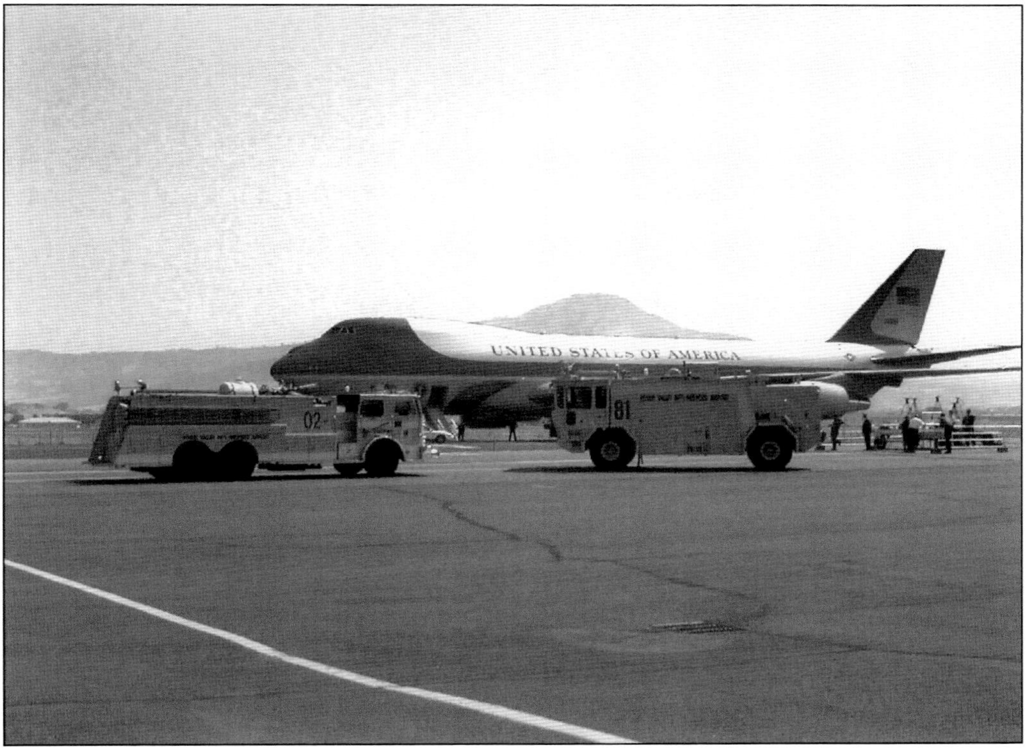

The Rogue Valley International-Medford Airport is now capable of accommodating just about any plane currently flying, including Air Force One, which has visited on a number of occasions carrying different presidents. (Rogue Valley International-Medford Airport.)

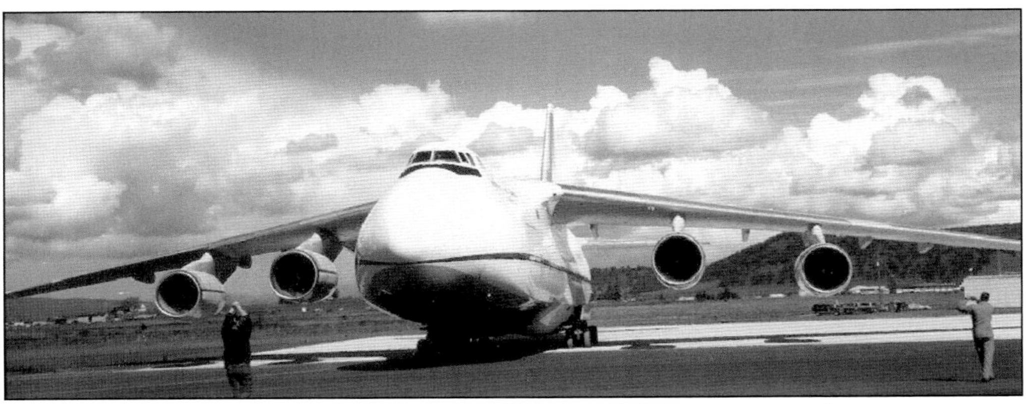

When Medford moved its airport from Newell Barber Field to its current location, it was to accommodate the new, giant eight-passenger transports of the day. Today Rogue Valley International-Medford Airport has grown up to handle just about any aircraft that flies, including this massive Russian-built Antonov AN-124. (Rogue Valley International-Medford Airport.)

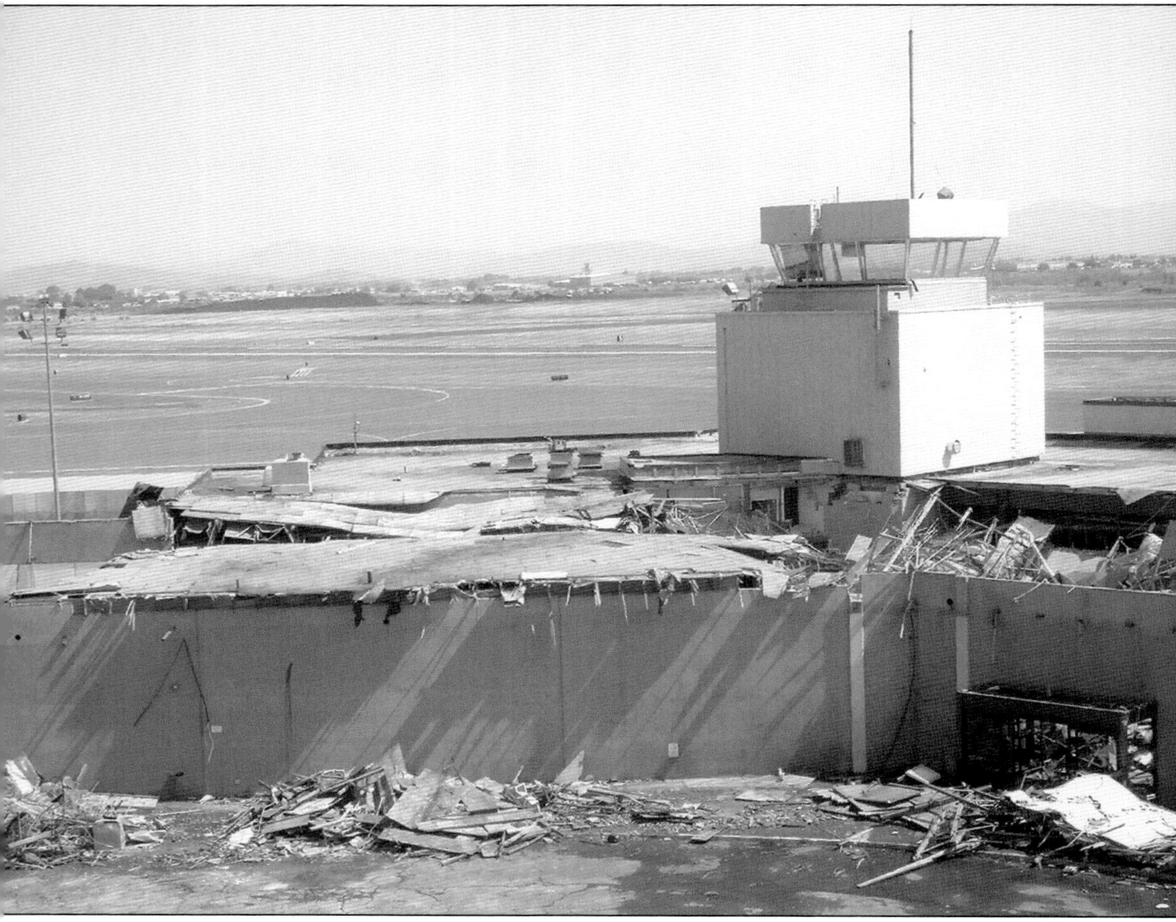

The Rogue Valley International-Medford airport continues to thrive and grow. In 2009, a third new terminal structure was completed and the second one was razed to accommodate expansion. (Rogue Valley International-Medford Airport.)

Seven

Epilogue: An Air-minded City

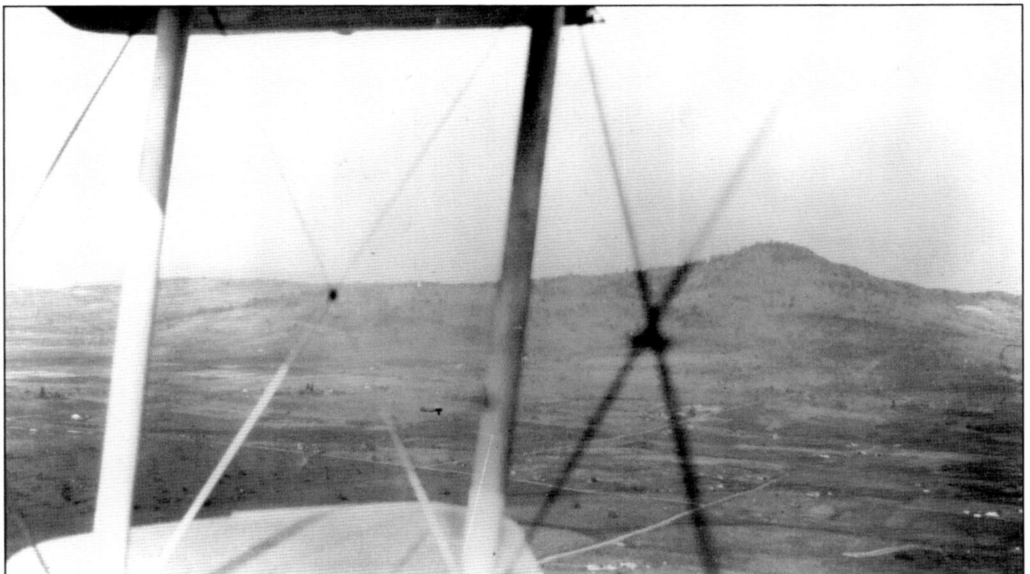

For more than 100 years, beginning with the age of the barnstormers, the people of southern Oregon have embraced aviation and worked to ensure its permanent place in the community. (SOHS No. 5046.)

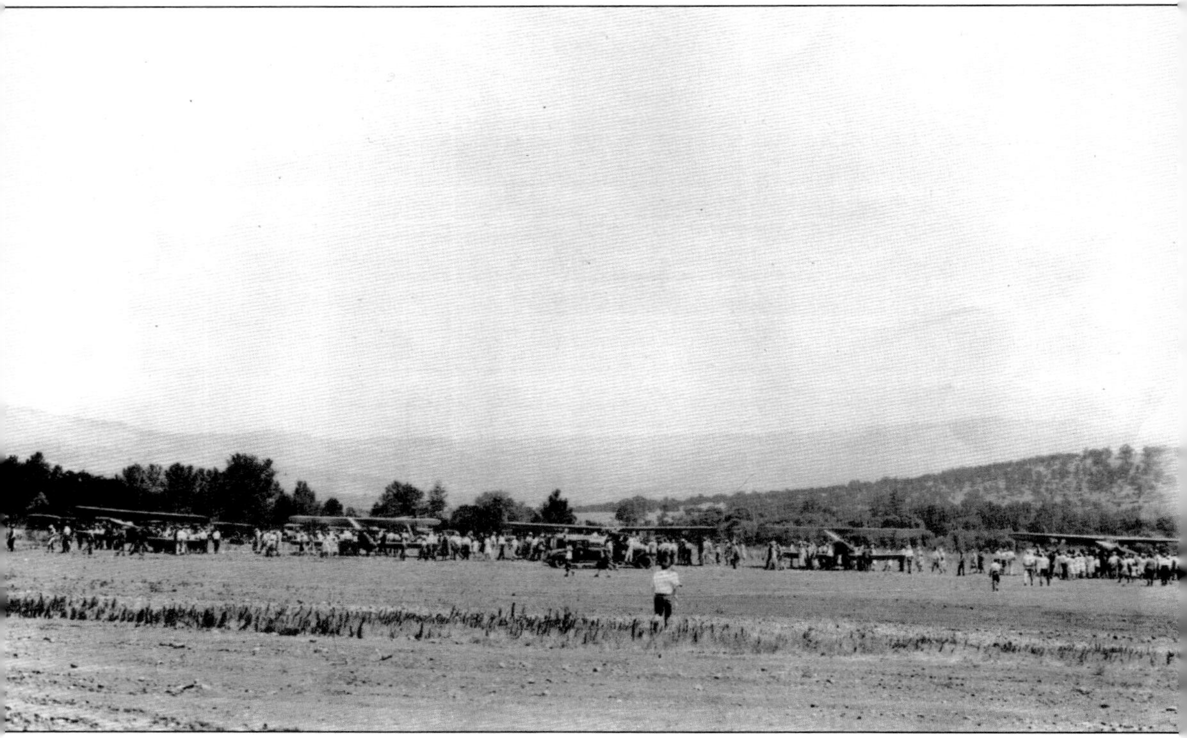

When the army air service and the forest service were searching for a location for forest fire patrols, it was the City of Medford that stepped forward and established the first municipally owned airfield in Oregon. (SOHS No. 9202.)

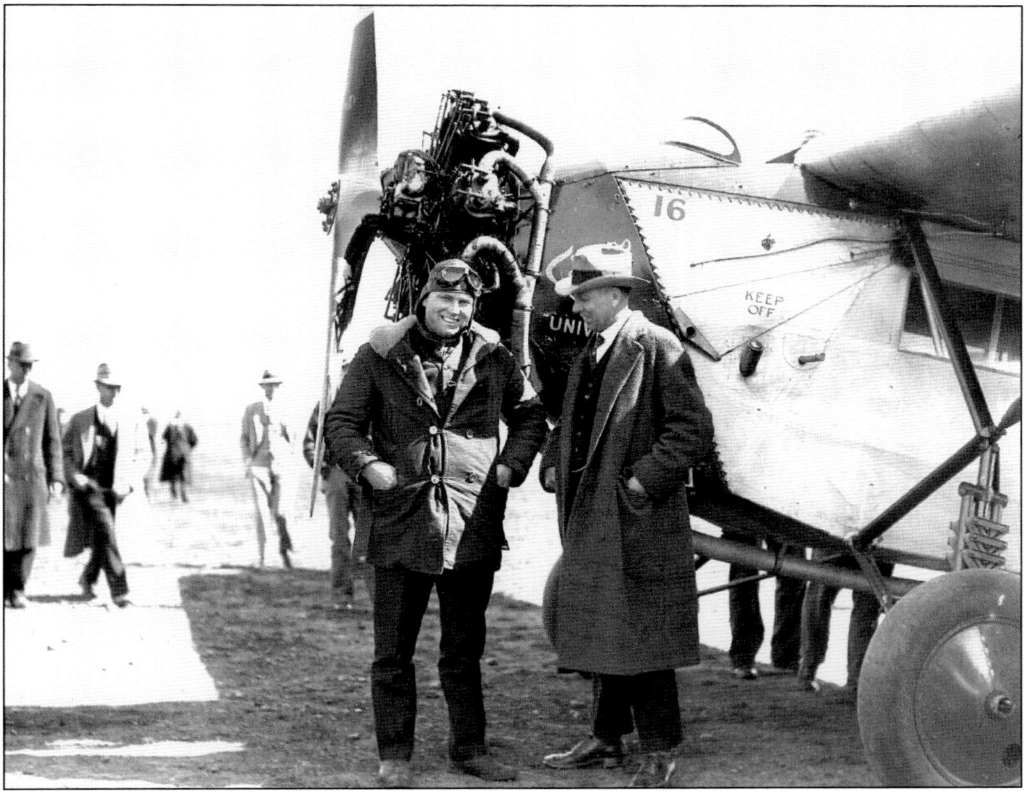

When Vern Gorst (right) selected Medford as a stop on the Pacific Air Transport route, many residents supported his efforts by investing in the fledgling company. Seely Hall, who would remain in its employ for his entire career, went so far as to sell shares out of his taxi. (Pearson Air Museum.)

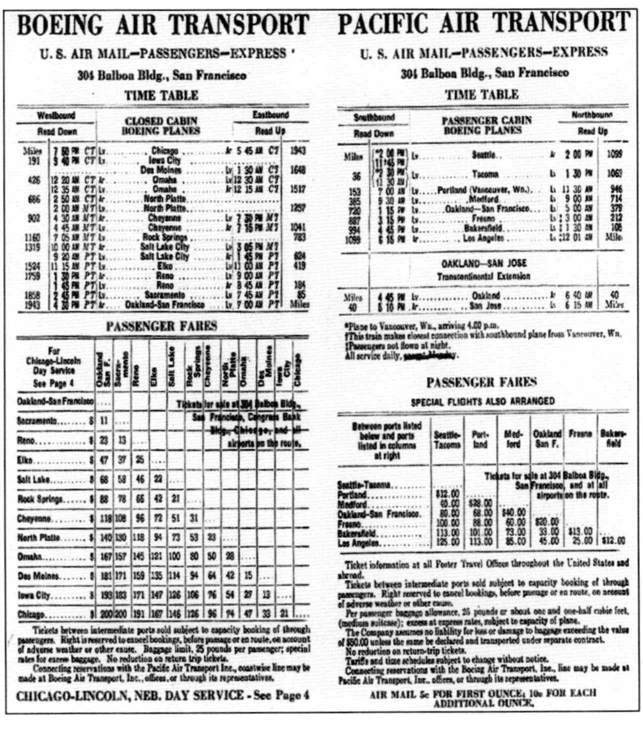

BOEING AIR TRANSPORT

U. S. AIR MAIL—PASSENGERS—EXPRESS

304 Balboa Bldg., San Francisco

TIME TABLE

Miles	Westbound Read Down		CLOSED CABIN BOEING PLANES		Eastbound Read Up	Miles
Miles	7 50 PM CT	Lv	Chicago	Ar	5 45 AM CT	1943
191	9 40 PM CT	Lv	Iowa City			
			Des Moines	Lv	1 30 AM CT	1648
426	12 20 AM CT	Ar	Omaha	Lv	12 30 AM CT	
	12 35 AM CT	Lv	Omaha	Ar	12 15 AM CT	1517
686	2 50 AM CT	Ar	North Platte			
	2 00 AM MT	Lv	North Platte			1257
902	4 30 AM MT	Ar	Cheyenne	Lv	7 30 PM MT	
	4 45 AM MT	Lv	Cheyenne	Ar	7 15 PM MT	1041
1160	7 05 AM MT	Lv	Rock Springs			783
1319	10 00 AM MT	Ar	Salt Lake City	Lv	3 05 PM MT	
	9 20 AM PT	Lv	Salt Lake City	Ar	1 45 PM PT	624
1524	11 15 AM PT	Lv	Elko	Lv	11 00 AM PT	419
1759	1 30 PM PT	Ar	Reno	Lv	9 00 AM PT	
	1 45 PM PT	Lv	Reno	Ar	8 45 AM PT	184
1858	2 45 PM PT	Lv	Sacramento	Lv	7 45 AM PT	85
1943	4 30 PM PT	Ar	Oakland-San Francisco	Lv	7 00 AM PT	Miles

PASSENGER FARES

For Chicago-Lincoln Day Service See Page 4	Oakland San F.	Sacra-mento	Reno	Elko	Salt Lake	Rock Springs	Cheyenne	North Platte	Omaha	Des Moines	Iowa City	Chicago
Oakland-San Francisco	....											
Sacramento........ $	11	....										
Reno............. $	23	13	....									
Elko............. $	47	37	25	....								
Salt Lake......... $	68	58	46	22	....							
Rock Springs...... $	88	78	66	42	21	....						
Cheyenne......... $	118	108	96	72	51	31	....					
North Platte...... $	140	130	118	94	73	53	23	....				
Omaha........... $	167	157	145	121	100	80	50	28	....			
Des Moines....... $	181	171	159	135	114	94	64	42	15	....		
Iowa City......... $	193	183	171	147	126	106	76	54	27	13	....	
Chicago.......... $	200	200	191	167	146	126	96	74	47	33	21	....

Tickets for sale at 304 Balboa Bldg., San Francisco, Congress Bank Bldg., Chicago, and all airports on the route.

Tickets between intermediate ports sold subject to capacity booking of through passengers. Right is reserved to cancel bookings, before passage or en route, on account of adverse weather or other cause. Baggage limit, 25 pounds per passenger; special rates for excess baggage. No reduction on return trip tickets.

Connecting reservations with the Pacific Air Transport Inc., coastwise line may be made at Boeing Air Transport, Inc., offices, or through its representatives.

CHICAGO-LINCOLN, NEB. DAY SERVICE - See Page 4

PACIFIC AIR TRANSPORT

U. S. AIR MAIL—PASSENGERS—EXPRESS

304 Balboa Bldg., San Francisco

TIME TABLE

Miles	Southbound Read Down		PASSENGER CABIN BOEING PLANES		Northbound Read Up	Miles
Miles	*2 00 PM 11 †45 PM	Lv	Seattle	Ar	2 00 PM	1099
36	*2 30 PM 11 30 AM	Lv	Tacoma	Lv	1 30 PM	1063
153	7 00 AM	Lv	Portland (Vancouver, Wn.)	Lv	11 30 AM	946
385	9 30 AM	Lv	Medford	Lv	9 00 AM	714
720	1 15 PM	Lv	Oakland—San Francisco	Lv	5 00 AM	379
887	3 15 PM	Lv	Fresno	Lv	‡3 00 AM	212
994	4 45 PM	Lv	Bakersfield	Lv	‡1 30 AM	105
1099	6 15 PM	Ar	Los Angeles	Lv	‡12 01 AM	Miles

OAKLAND—SAN JOSE
Transcontinental Extension

Miles						Miles
Miles	4 45 PM	Lv	Oakland	Ar	6 40 AM	40
40	5 10 PM	Ar	San Jose	Lv	6 15 AM	Miles

*Plane to Vancouver, Wn., arriving 4.00 p.m.
†This train makes closest connection with southbound plane from Vancouver, Wn.
‡Passengers not flown at night.
All service daily, .

PASSENGER FARES

SPECIAL FLIGHTS ALSO ARRANGED

Between ports listed below and ports listed in columns at right	Seattle-Tacoma	Port-land	Med-ford	Oakland San F.	Fresno	Bakers-field
Seattle-Tacoma........	......					
Portland..............	$12.00	......				
Medford..............	40.00	$28.00	......			
Oakland-San Francisco.	80.00	68.00	$40.00	......		
Fresno..............	100.00	88.00	60.00	$20.00	......	
Bakersfield...........	113.00	101.00	73.00	33.00	$13.00	...
Los Angeles..........	125.00	113.00	85.00	45.00	25.00	$12.00

Tickets for sale at 304 Balboa Bldg., San Francisco, and at all airports on the route.

Ticket information at all Foster Travel Offices throughout the United States and abroad.

Tickets between intermediate ports sold subject to capacity booking of through passengers. Right reserved to cancel bookings, before passage or en route, on account of adverse weather or other cause.

Per passenger baggage allowance, 25 pounds or about one and one-half cubic feet, (medium suitcase); excess at express rates, subject to capacity of plane.

The Company assumes no liability for loss or damage to baggage exceeding the value of $50.00 unless the same be declared and transported under separate contract.

No reduction on return-trip tickets.

Tariffs and time schedules subject to change without notice.

Connecting reservations with the Boeing Air Transport, Inc., line may be made at Pacific Air Transport, Inc., offices, or through its representatives.

AIR MAIL 5c FOR FIRST OUNCE; 10c FOR EACH ADDITIONAL OUNCE.

Pictured is a timetable for Pacific Air Transport. PAT was southern Oregon's first commercial carrier, and its corporate heir, United Airlines, continues to service Medford, along with Horizon Air and Delta Air Lines. (SOHS Manuscript.)

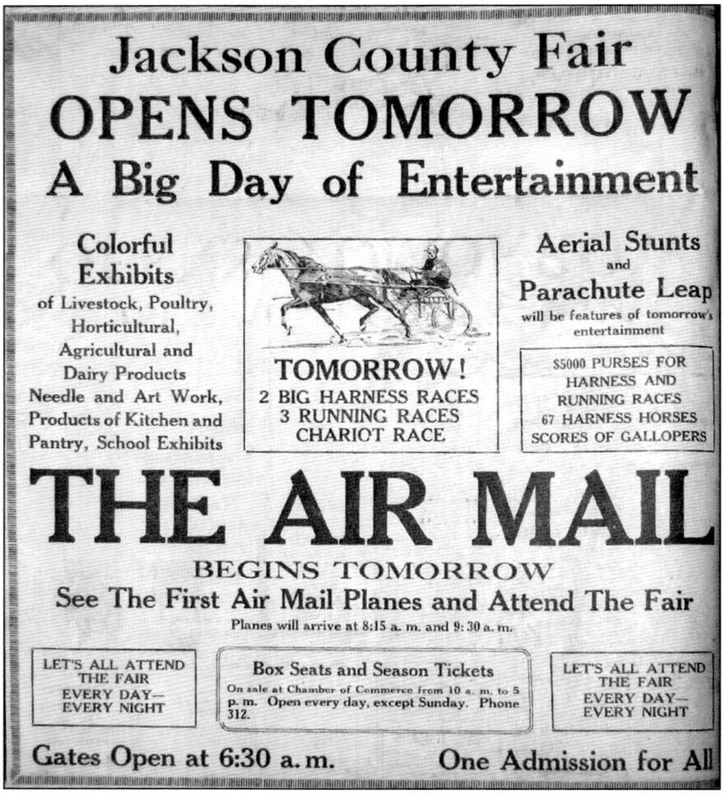

In a show of support, Jackson County scheduled the opening of the county fair to coincide with the arrival of the first airmail, September 15, 1926. (*Medford Mail Tribune.*)

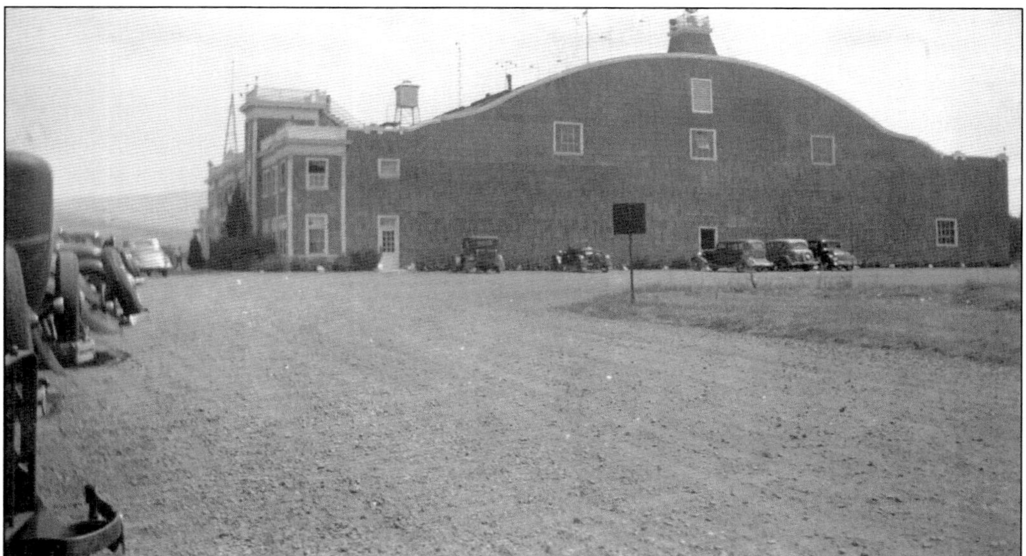

When Medford was threatened with the loss, through obsolescence, of its airfield, citizens overwhelmingly voted to support a $120,000 bond issue to build a new airport. (SOHS No. 20712.)

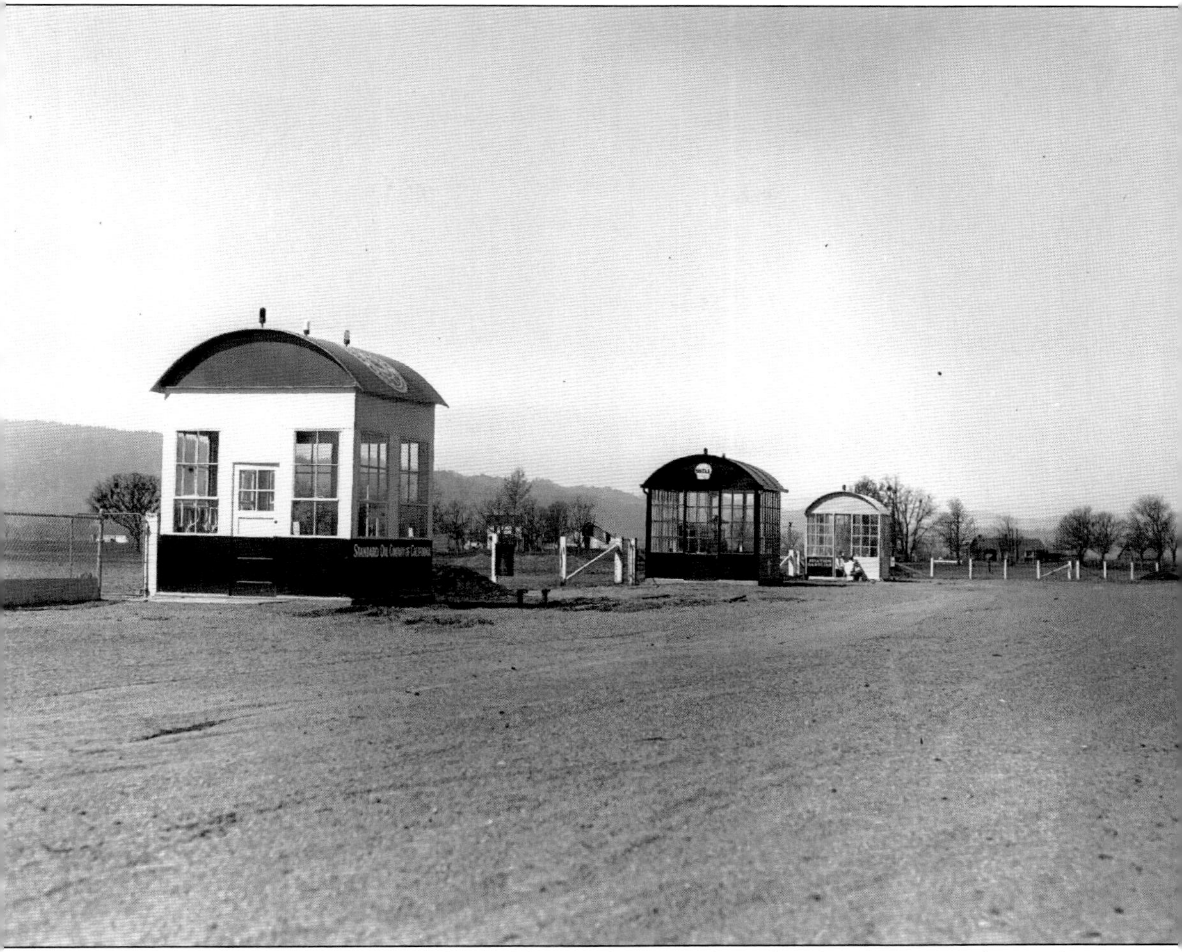

Medford's voters replaced Newell Barber Field with a large, modern class A airport, complete with weather facilities, a department of commerce radio facility, and lights for night flying. There was also fuel service from several suppliers, making the new Medford Airport one of the best on the West Coast. (SOHS No. 19188.)

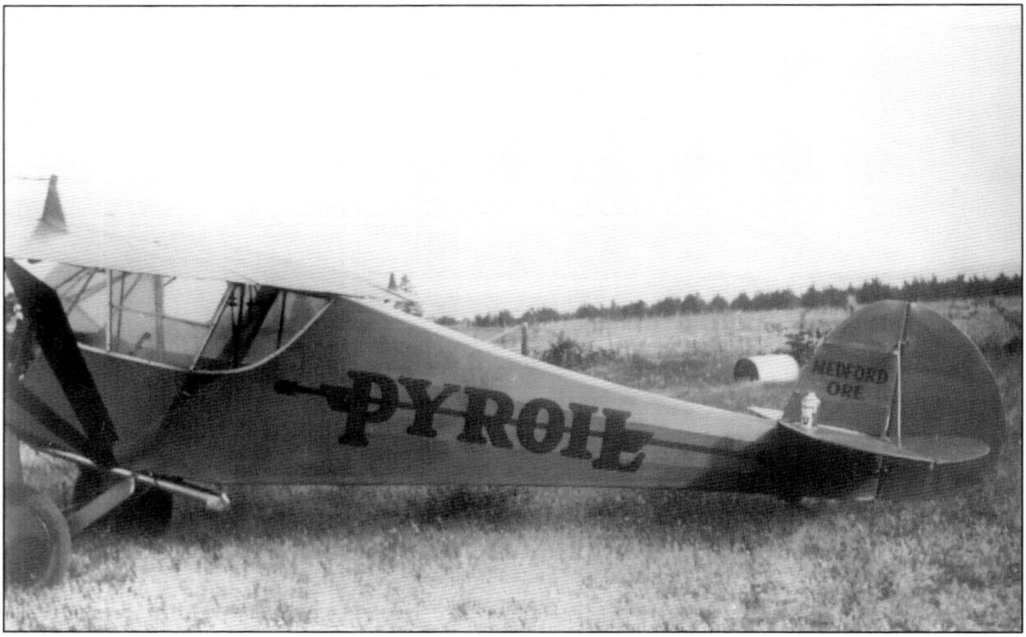

It was not just commercial aviation that propelled Medford to the forefront of Oregon aviation. General aviation's roots with Hall and Hart's *Mayfly* have grown into a major regional hub for aviation. Aubrey Sander advertised Pyroil aviation products on his plane. (SOHS No. 12730.)

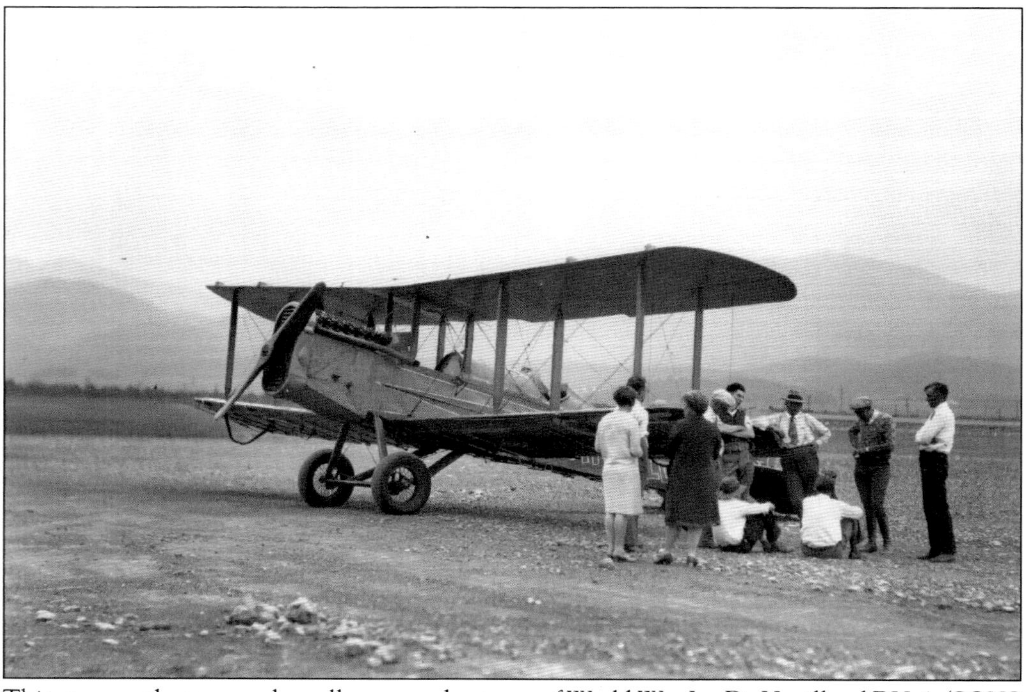

This group gathers around a well-preserved veteran of World War I, a De Havilland DH-4. (SOHS No. 19207.)

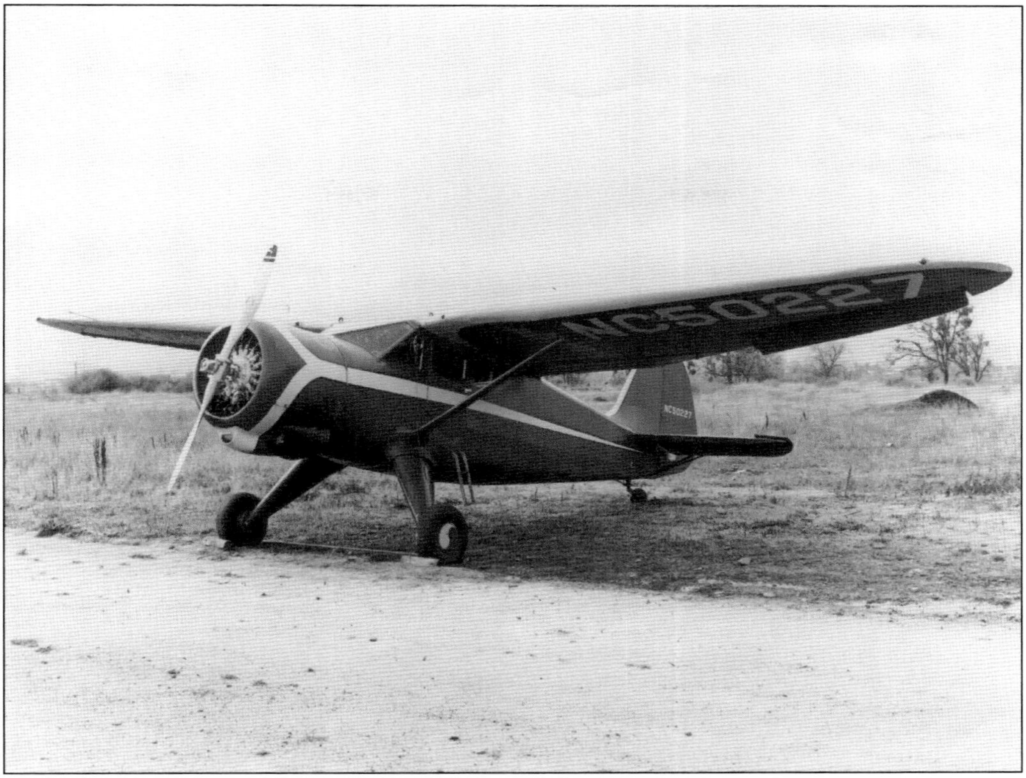

Medford's contributions also include the establishment of Mercy Flights, which still provides air and ground ambulance service to the entire Rogue Valley. One of its smaller planes, a Stinson Reliant, sits, ready for action. (SOHS No. 4556.)

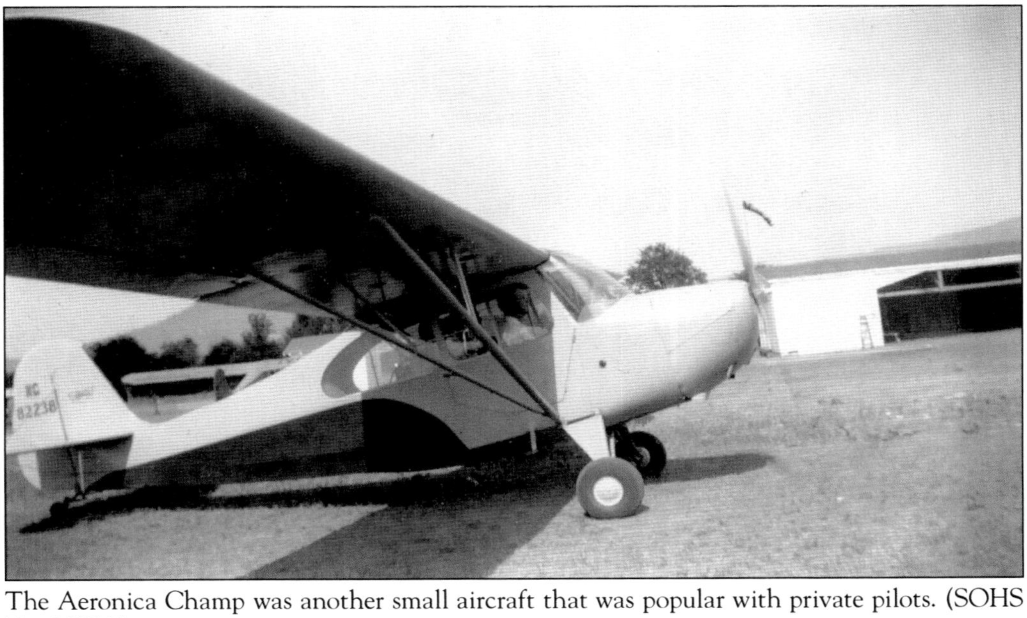

The Aeronica Champ was another small aircraft that was popular with private pilots. (SOHS No. 20700.)

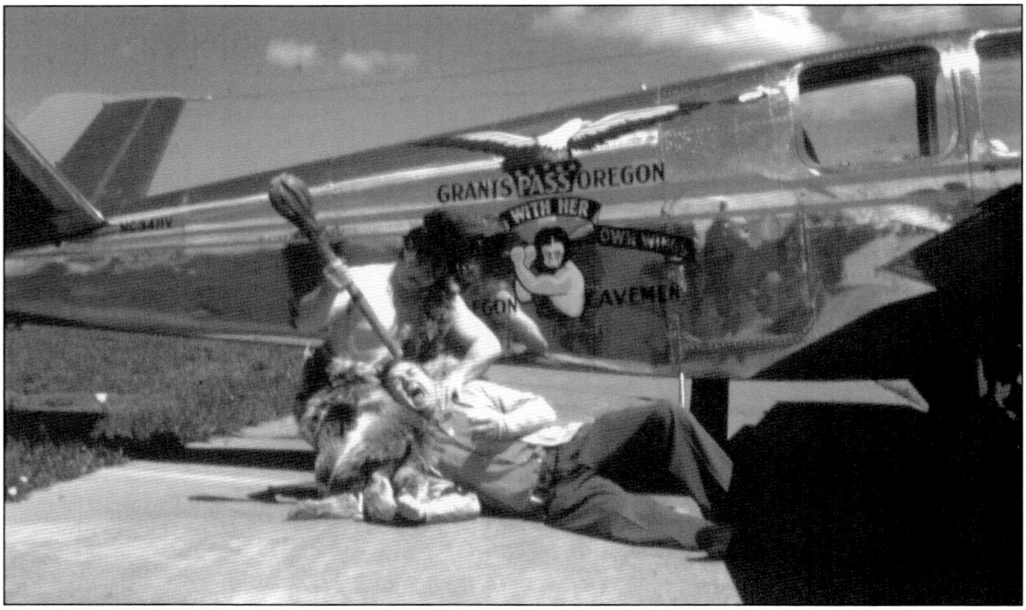

In 1949, the chamber of commerce of Portland, Oregon, organized a tour to promote the Rose City. Their goal was to fly from Portland, Oregon, to Portland, Maine. A contingent of southern Oregon pilots representing Medford, Grants Pass, and Coos Bay decided to join the tour. The Grants Pass cavemen brought with them their usual Stone Age antics, this particular victim being Medford pilot Dick Alley. (Author's collection.)

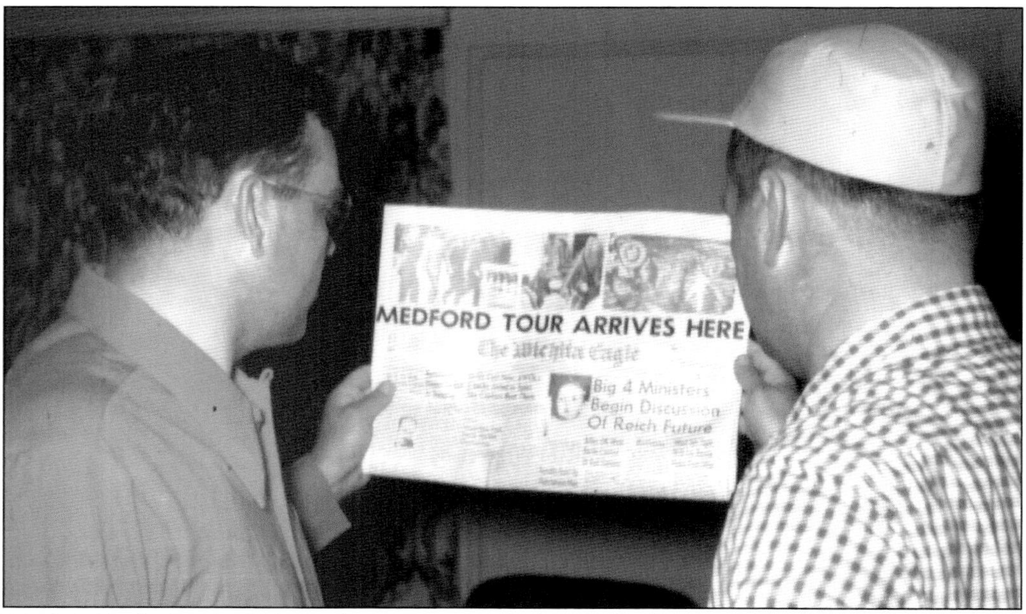

The Medford pilots had planned to meet up with the Portland tour in Wichita, Kansas. Arriving well ahead of the main body of planes, Dick Alley found time to have a mock newspaper printed, hailing the arrival of the Medford Air Tour. Organizers were not happy with the stunt, but the separate Medford identity caught on for the remainder of the tour. Bob Alley (right) examines his brother's handiwork. (Author's collection.)

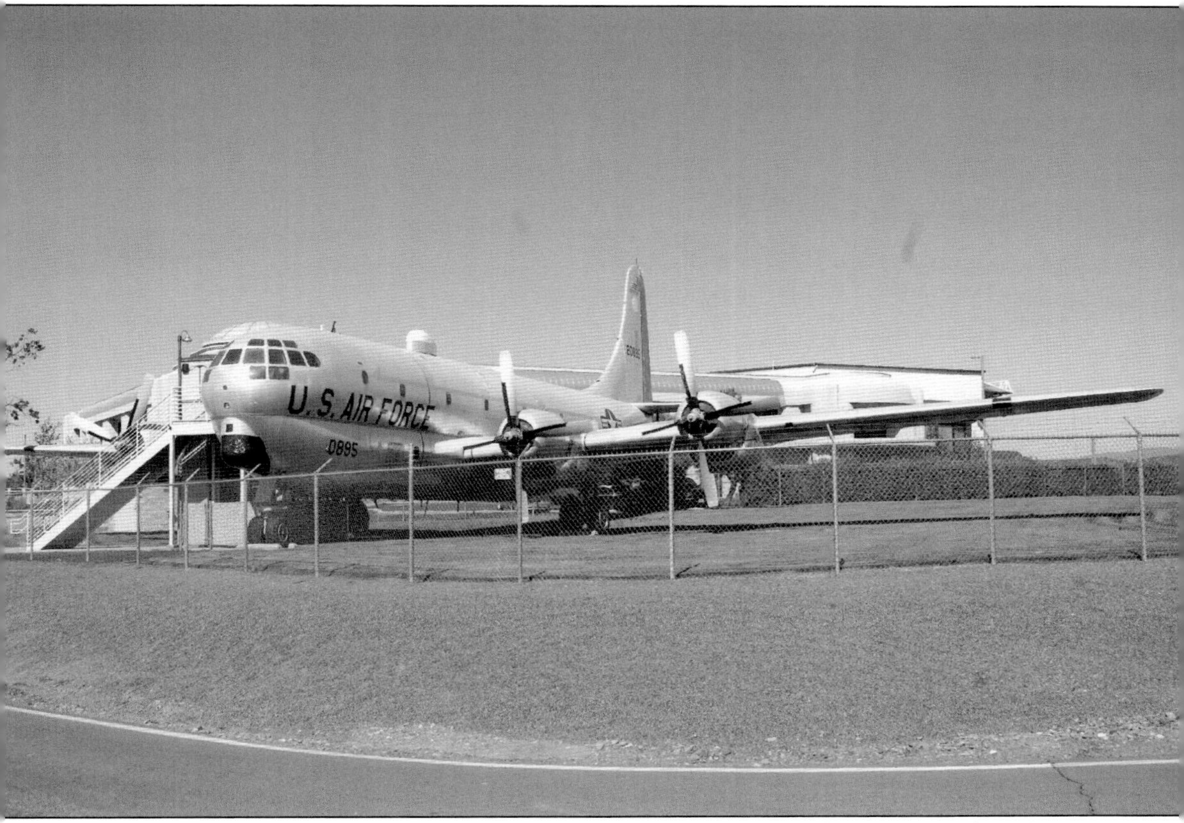

One little-known attraction at the Rogue Valley International-Medford Airport, is the KC-97 Stratotanker. Built in 1952, this plane saw long service with the Strategic Air Command before arriving in Medford for retirement in 1990. Eager volunteers carefully restored the plane and made it accessible according to the Americans with Disabilities Act (ADA) standards. It is now a popular attraction and rental facility. (Author's collection.)

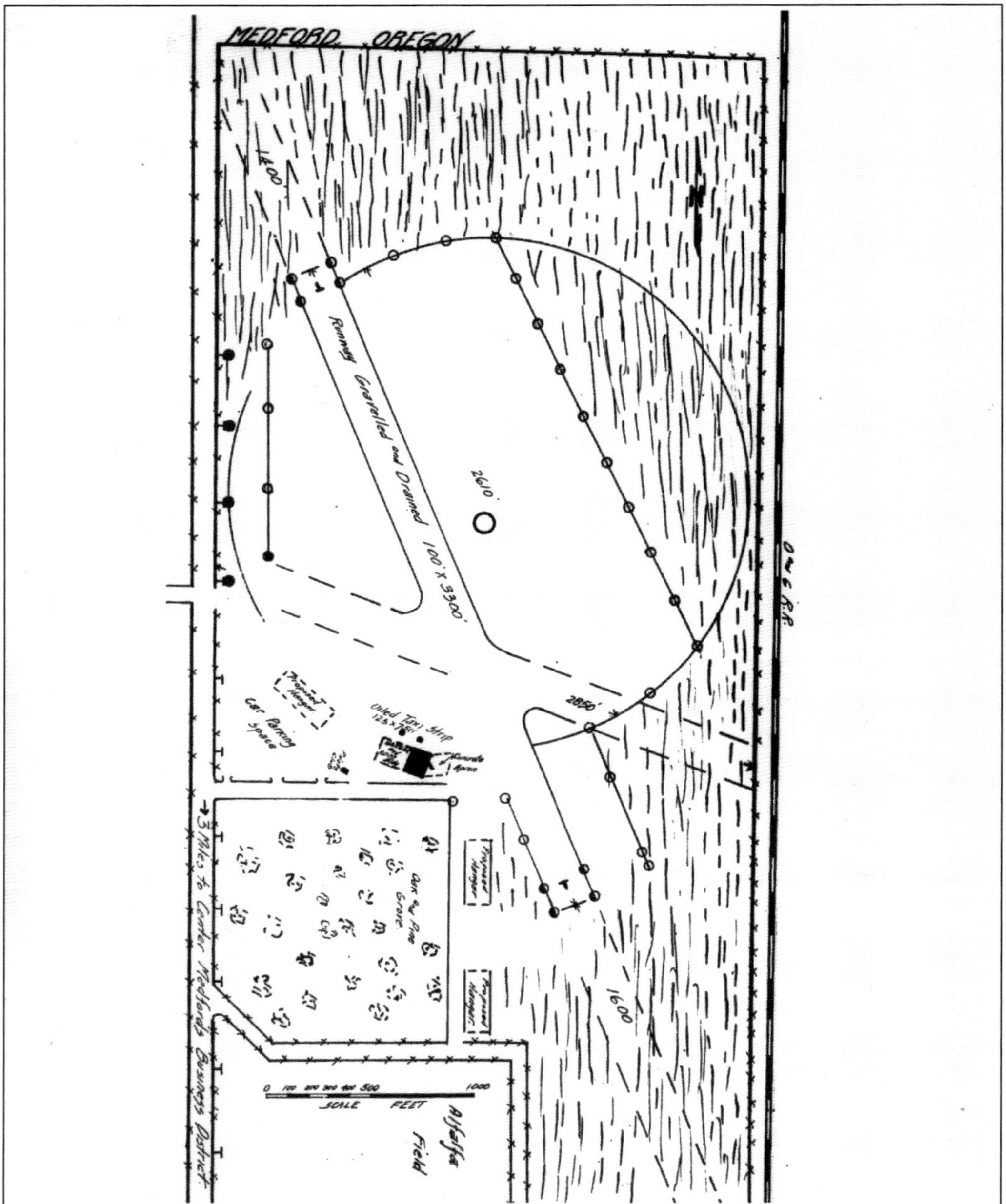

This early map of the Medford Municipal Airport was used to illustrate just how modern and advanced the new field, dedicated in 1930, was in relation to Newell Barber Field. (SOHS No. 15626.)

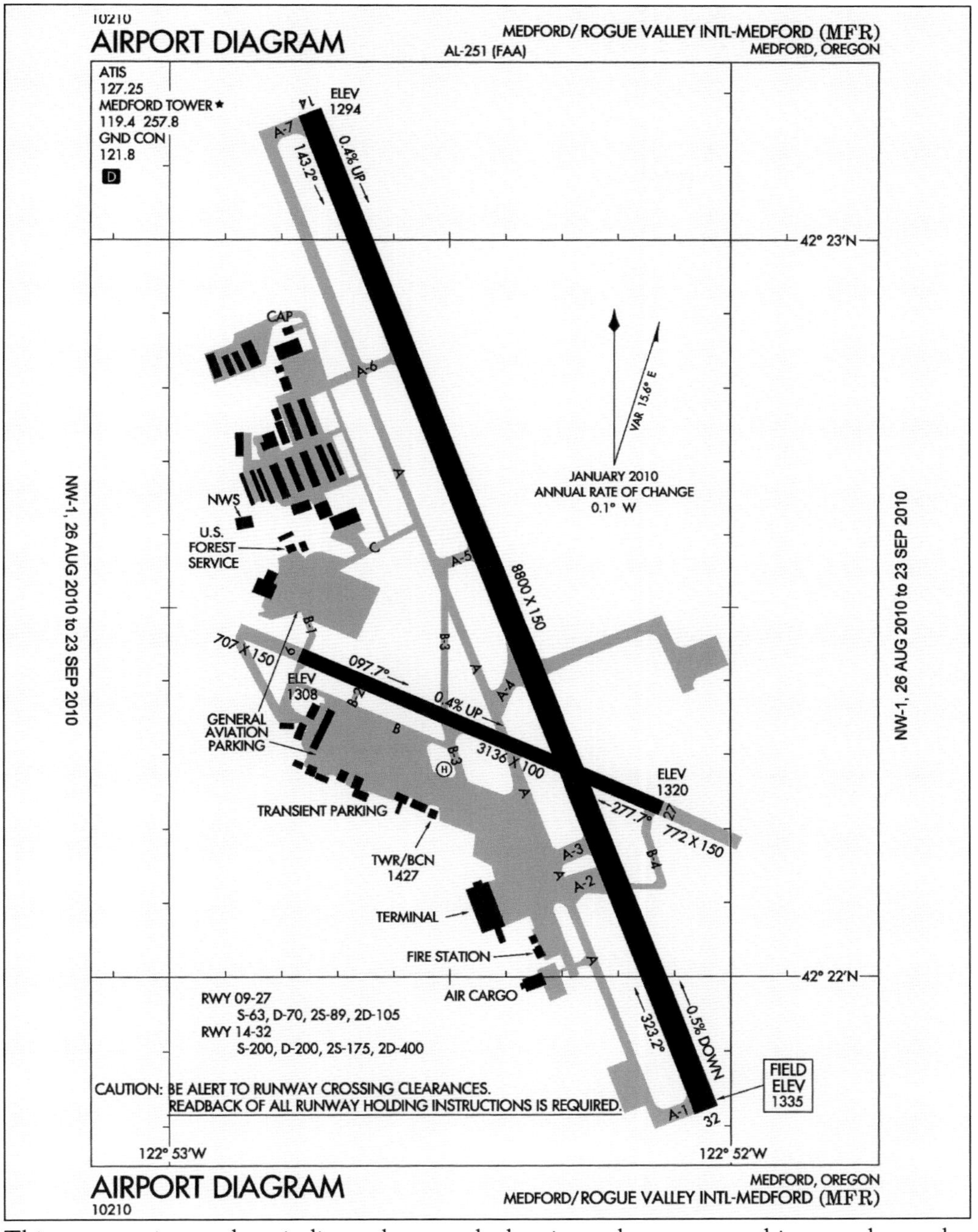

This current airport chart indicates how much the airport has grown and improved over the course of the 80 years it has occupied the Biddle Road location. (Federal Aviation Administration/ Government Printing Office.)

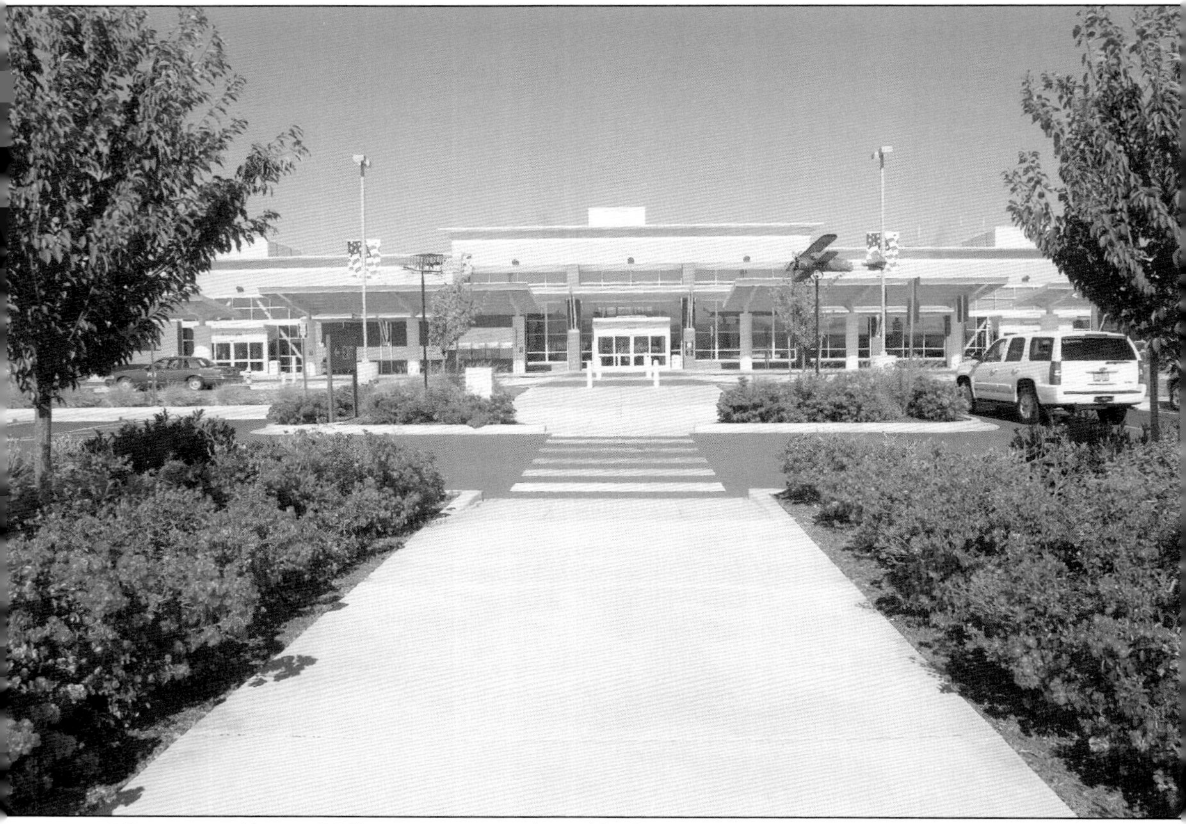

The newest Rogue Valley International-Medford Airport terminal was dedicated in 2009 and is the third facility on this site since the airport moved to this location more than 80 years ago. (Author's collection.)

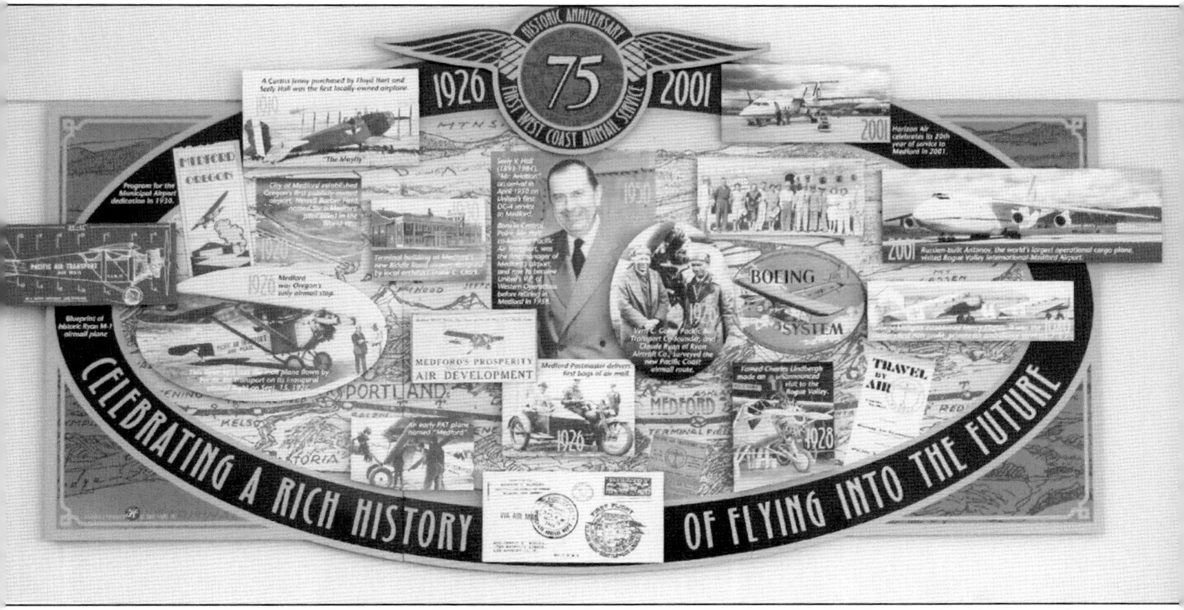

In 2001, Medford celebrated the 75th anniversary of West Coast airmail service. Today the grand mural installed that day is on display in the main terminal. Now, almost a decade later, Medford can look forward to its centennial as Oregon's oldest "Air-Minded City." (Author's collection.)